Unsupervised and Semi-Supervised Learning

Springer's Unsupervised and Semi-Supervised Learning book series covers the latest theoretical and practical developments in unsupervised and semi-supervised learning. Titles – including monographs, contributed works, professional books, and textbooks – tackle various issues surrounding the proliferation of massive amounts of unlabeled data in many application domains and how unsupervised learning algorithms can automatically discover interesting and useful patterns in such data. The books discuss how these algorithms have found numerous applications including pattern recognition, market basket analysis, web mining, social network analysis, information retrieval, recommender systems, market research, intrusion detection, and fraud detection. Books also discuss semi-supervised algorithms, which can make use of both labeled and unlabeled data and can be useful in application domains where unlabeled data is abundant, yet it is possible to obtain a small amount of labeled data.

Topics of interest in include:

- Unsupervised/Semi-Supervised Discretization
- Unsupervised/Semi-Supervised Feature Extraction
- Unsupervised/Semi-Supervised Feature Selection
- Association Rule Learning
- Semi-Supervised Classification
- Semi-Supervised Regression
- Unsupervised/Semi-Supervised Clustering
- Unsupervised/Semi-Supervised Anomaly/Novelty/Outlier Detection
- Evaluation of Unsupervised/Semi-Supervised Learning Algorithms
- Applications of Unsupervised/Semi-Supervised Learning

While the series focuses on unsupervised and semi-supervised learning, outstanding contributions in the field of supervised learning will also be considered. The intended audience includes students, researchers, and practitioners.

More information about this series at http://www.springer.com/series/15892

Frédéric Ros • Serge Guillaume
Editors

Sampling Techniques for Supervised or Unsupervised Tasks

Springer

Editors
Frédéric Ros
PRISME Laboratory
University of Orléans
Orléans, France

Serge Guillaume
UMR ITAP
Irstea, Montpellier, France

ISSN 2522-848X ISSN 2522-8498 (electronic)
Unsupervised and Semi-Supervised Learning
ISBN 978-3-030-29348-2 ISBN 978-3-030-29349-9 (eBook)
https://doi.org/10.1007/978-3-030-29349-9

This Springer imprint is published by the registered company Springer Nature Switzerland AG.
The registered company address is: Gewerbestrasse 11, 6330 Cham, Switzerland

In science the difficulty is not to have ideas,
but it is to make them work.

Carlo Rovelli

Preface

The big data era is characterized by an exponential increase of volume, velocity, and variety of the available data. This raises various challenges for the data processing tasks that require not only to store, organize, and provide a granular access to the data, but also to clean and analyze their content. Data mining is an important part of the data analysis as it contributes to knowledge discovery by identifying understandable patterns in data.

The potential of these techniques has already attracted interest from users and decision-makers as well as from researchers of fields such as statistics or computer science. They proved efficient in various domains, for example, medicine and astronomy. However, there is still a growing demand for efficient and scalable solutions to handle some task, clustering is a representative example, with limited resources both in time and memory space.

This opened several and complementary ways of research. The first one aims to take advantage of the new capabilities of hardware components. Famous examples are parallel computing, use of graphics processing unit (GPU) for fast calculations. The second one deals with the algorithms themselves: how to make them faster, more tractable? But what is gained by the optimization procedures is likely to be compensated by the need to deal with more complex data, for example, shapes and densities in clustering, or with an increased volume. In this context, sampling appears as a third option. Working with a representative sample and optimized algorithms with an implementation that takes advantage of the hardware structure may result successful.

Defining a sample that behaves like the whole data set is a quite long-standing issue in data management. It has received fresh interest with the challenge of big data. Sampling may be applied to instances (or items), or features, with the goal of unsupervised (e.g., clustering) or supervised tasks (e.g., classification). This book provides the reader with an up-to-date review of some important sampling techniques. It is organized in seven chapters.

Sampling has strong statistical roots, even if this is sometimes forgotten in the machine learning community. In Chap. 1, Guillaume Chauvet, research professor from ENSAI/IRMAR (France), describes the framework of a sampling design, and the associated estimators and measures of accuracy. Then, he introduces several sampling algorithms that can be used for sample selection, with discussion of their practical and theoretical pros and cons.

Chapter 2, by Dan Feldman from University of Haifa (Israel), is dedicated to an updated review on coresets. The author participated in the main developments of this notion introduced in 2003. A coreset is a summary of the input, where every possible query (usually from an infinite set of queries) has approximately the same answer on both data sets. Generic techniques enable efficient coreset maintenance of streaming, distributed and dynamic data. An approximation to the optimal solution of the original input can then be maintained via the coreset.

The challenge is to design coresets with provable trade-off between their size and approximation error. The objective of this survey is to introduce the state of the art of these techniques.

Chapter 3, by Serge Guillaume from Irstea (France) and Frédéric Ros from Université d'Orléans (France), introduces a family of unsupervised sampling algorithms based on distance and density concepts. Once the common characteristics are described, the three algorithms are detailed and compared. An open source implementation of the proposed algorithms is available. This chapter also includes a state-of-the-art section about unsupervised sampling.

Chapter 4, by Frédéric Ros and Serge Guillaume, reviews the data reduction problem for instance and feature selection methods in the context of supervised classification. In the first part, instance and feature selections are studied separately. As instance and feature selection are not independent, algorithms dealing with simultaneous selection are then presented. To provide a comprehensive and tractable view of this field, the strategy was to start from the fundamental and original contributions go towards state-of-the-art algorithms, paying particular attention to large-scale selections. Detailed pseudo codes of representative algorithms are given to consolidate the whole.

Chapter 5, by Nicolas Tremblay from CNRS, GIPSA-lab (France) and Andreas Loukas from Ecole Polytechnique Fédérale de Lausanne (Switzerland), reviews sampling methods that were specifically designed for spectral clustering: a family of well-known unsupervised learning algorithms. In a nutshell, spectral clustering techniques do not attempt to cluster data in their native domain of representation and are the combination of three main steps: (1) transform the data by constructing a (usually sparse) similarity graph, (2) compute the principal eigenvectors of its Laplacian, and (3) cluster (with k-means for instance) the transformed data in the spectral domain. One of the advantages of this framework is to be able to separate non-convex clusters. The main drawback of these methods is their computational cost. This chapter reviews available sampling methods that are specifically designed to accelerate one or more of these three steps, focusing on their approximation guarantees when available and discusses practical merits and limitations.

The last two chapters illustrate one of the new challenges of sampling: dealing with complex objects, instead of traditional vectors, and dynamic data flows. In Chap. 6, by Ali Idarrou and H. Douzi, from the University of Agadir (Morocco), deals with managing streams of unstructured data, such as those provided by social networks. In the first part, the authors review the concepts of sliding window and reservoir techniques that are useful for managing data streams. The processing of dynamic complex data, coming from heterogeneous sources and likely to include some inconsistency and uncertainty, raises many problems. Conventional sampling tools and techniques used with simple vector data may give poor results. The second part of this chapter reviews the main techniques of integration and data fusion. The goal is to take advantage of the complementarity and redundancy of the different information sources to get a more robust information.

Chapter 7, by Javier Calle et al. from Carlos III University of Madrid (Spain), proposes methods to handle huge dynamic graphs using Ant Colony Optimization algorithms (ACO). The main focus is on the path search problem: while many algorithms try to find the best path, either the shortest or the cheapest, the goal with huge dynamic graphs is to privilege response time. The algorithms benefit from the storage and quick access to the data provided by the database technology. The authors show that their novel metaheuristic method of the ACO is good enough to avoid the expiration of the validity of the solution provided in most cases.

This book comes from the machine learning community, and it is suited to anyone, practitioner, student, or teacher, who wants to become more familiar with the sampling techniques as they can be helpful to tackle the big data challenge. The chapters are independent and complementary, from theoretical concepts to practical implementations. They can be read in any order. Have fun in the sampling world!

Orléans, France — Frédéric Ros
Montpellier, France — Serge Guillaume

Contents

About the Editors

Frédéric Ros has an engineering degree in Microelectronics and Automation, a master's in Robotics from Montpellier University, and a Ph.D. from ENGREF (Ecole Nationale du Genie Rural des Eaux et Forets), Paris. He began his career in 1991 as a research scientist working on the field of image analysis for robotics and artificial systems from IRSTEA. He managed the vision activity in GEMALTO for 14 years which is the world leader in the smart card industry. He was particularly involved in applied developments (related to machine vision, AI, data analysis, fuzzy logic, and supervised classification algorithms) with the aim of providing adaptive and self-tuning systems corresponding to the growing complexity of industrial processes and especially multidisciplinary interactions. Professor at Orleans University, he has been an associate researcher at PRISME laboratory (signal and image processing) for 10 years, focusing his research on machine learning algorithms in the big data context. In parallel, he has head an innovation park for 11 years and recently designed and structured a digital start-up incubator.

Serge Guillaume got his Ph.D. in Computer Science in 2001 and his habilitation thesis (HDR) in 2005, both from Toulouse University. He has been working for Irstea, a public National Research Institute of Science and Technology for Environment and Agriculture, for 30 years. He has been involved in fuzzy systems design and optimization for 20 years. He is now also interested in the design of decision support systems based on spatial data, and this includes spatial modeling, for example, zone delineation and multicriteria decision-making. His work is implemented as two open source software, FisPro and GeoFIS.

About the Editors

Chapter 1
Introduction to Sampling Techniques

Guillaume Chauvet

1.1 Introduction

The theory of sample surveys aims at selecting a sample of units, so as to represent a larger population. Although such techniques are now of routine use, in particular by national statistical agencies, the legitimacy of this approach has long been debated. The introduction of the representative method dates back to [42], who proposes to produce estimations by using a non-random controlled sample of municipalities and individuals, rather than a census. But this is truly with [46] that the basics of modern survey sampling are established. Neyman proposes a rigorous setup for random surveys, laying the foundations of probabilistic yet controlled surveys, which enable to statistically control the accuracy of estimators.

During decades, many of the available data sets emerged from surveys, the bigger ones being produced by the national statistical agencies. These samples of reasonably large size (a few thousand up to a few tens of thousands of units) aimed at shortening the delay of production of statistical information, with limited cost. For the past 10 years, we have been facing a completely new situation with a burst of the volume of the generated data. It is estimated that 200,000 sms are sent every second worldwide, Facebook users generate four million likes every minute, and in 2020, each person will produce 1.7 MB of data per second. These data cannot be saved and treated comprehensively. The theory of sample surveys is therefore needed to select large but tractable samples, sometimes processed in real time, to faithfully represent these huge masses of data.

More than ever, we need to control the statistical properties of the estimators produced. The purpose of this chapter is therefore to provide an introduction to the

G. Chauvet (✉)
ENSAI/IRMAR, Bruz, France
e-mail: chauvet@ensai.fr

F. Ros, S. Guillaume (eds.), *Sampling Techniques for Supervised or Unsupervised Tasks*, Unsupervised and Semi-Supervised Learning,
https://doi.org/10.1007/978-3-030-29349-9_1

estimation framework of survey sampling, by reviewing some of the classical sampling methods. For the sake of simplicity, we focus on direct sampling procedures, which may be seen as the basic bricks needed to build a sampling design in practice. For the processed sampling methods, their statistical properties are discussed. For the reader interested by a more complete presentation of survey sampling tools, see [1, 16, 52, 58] and [28], for example.

This chapter is organized as follows: In Sect. 1.2, we present the survey sampling framework, and describe the classical Horvitz–Thompson estimator and the associated measures of accuracy. In Sect. 1.3, we focus on simple random sampling, for which we describe the statistical properties and give a number of possible sampling algorithms. In Sect. 1.4, we describe a number of classical unequal probability sampling algorithms, and we discuss their statistical properties in Sect. 1.5. We conclude in Sect. 1.6.

1.2 Notation and Estimation

In this section, we define our main notation and introduce the Horvitz–Thompson estimator which is of routine use in survey sampling. We then discuss variance computation and variance estimation for the Horvitz–Thompson estimator.

1.2.1 Notation

We are interested in a finite population U_N consisting of N statistical units, which are supposed to be easily identified by a label. Therefore, it is common practice to make no distinction between a unit and its label, and we simply write the population as

$$U_N = \{1, \ldots, N\} \tag{1.1}$$

We are interested in some quantitative variable of interest y, taking the value y_k on unit k.

We suppose that the population of interest U_N is embedded into a nested sequence $\{U_N\}$ of finite populations with increasing sizes, and all limiting processes will be taken as $N \to \infty$. This is essentially the asymptotic framework of [41], which is often used to study the asymptotic properties of a sampling design and of the assorted estimators. Also, this is a natural framework if we are interested in a population which is growing over time, for example, if we wish to select a sample in a data stream.

A without-replacement sampling design $p_N(\cdot)$ is a probability distribution on the subsets in U_N, namely

$$\forall s \subset U_N \quad p_N(s) \geq 0 \text{ and } \sum_{s \subset U_N} p_N(s) = 1 \tag{1.2}$$

It enables selecting the random sample S_N of units used for estimation, in the sense that $Pr(S_N = s) = p_N(s)$. Once the sampling design is defined, we need to choose a sampling algorithm, which is an effective procedure for the selection of the sample. For a given sampling design, there is usually a variety of sampling algorithms possible [58], see Sect. 1.3 for an illustration on simple random sampling.

The quantity $\pi_{k|N} \equiv Pr(k \in S_N)$ for unit k to be selected is called the first-order inclusion probability. The $\pi_{k|N}$'s are involved in the computation of estimators, and their sum

$$\sum_{k \in U_N} \pi_{k|N} \equiv n \tag{1.3}$$

gives the average sample size. The probability $\pi_{kl|N} \equiv Pr(k, l \in S_N)$ for units k and l to be jointly selected in S_N is called the second-order inclusion probability. The $\pi_{kl|N}$'s are involved in the computation of the variance of estimators. For a given set of first-order inclusion probabilities $\pi_{k|N}$, $k \in U_N$, the second-order inclusion probabilities depend on the design used for the selection of the sample.

1.2.2 Horvitz–Thompson Estimator

The Horvitz and Thompson [39] estimator (HT) of the total $t_{yN} = \sum_{k \in U_N} y_k$ is

$$\hat{t}_{y\pi} = \sum_{k \in S_N} \frac{y_k}{\pi_{k|N}} = \sum_{k \in U_N} I_{kN} \frac{y_k}{\pi_{k|N}} \tag{1.4}$$

with I_{kN} the sample membership indicator of unit k. We note

$$I_N = (I_{1N}, \ldots, I_{kN}, \ldots, I_{NN}) \tag{1.5}$$

the vector of sample membership indicators. If all the $\pi_{k|N}$'s are positive, which is assumed in the rest of the paper, there is no selection bias. In such case, the HT-estimator is design-unbiased for t_{yN}, i.e., unbiased under the randomization associated with the sampling design. It is remarkable that this property comes completely model-free. It holds for any variable of interest, without requiring any model assumptions.

There is no severe loss of generality in focusing on the total t_{yN}, since many other parameters of interest can be written as smooth functions of totals. Such parameters are therefore easily estimated in a plug-in principle once an estimator of a total is available, see [20]. For example, the population mean is $\mu_{yN} = N^{-1} \sum_{k \in U_N} y_k$, and is estimated by

$$\hat{\mu}_{y\pi} = \frac{\hat{t}_{y\pi}}{\hat{N}_\pi} \tag{1.6}$$

where $\hat{N}_\pi = \sum_{k \in S_N} \frac{1}{\pi_{k|N}}$ is the HT-estimator of the population size N. Similarly, the population distribution function for some real number t is $F_{yN}(t) = N^{-1} \sum_{k \in U_N} 1(y_k \leq t)$, with $1(\cdot)$ the indicator function. The plug-in estimator of $F_{yN}(t)$ is

$$\hat{F}_{y\pi} = \frac{1}{\hat{N}_\pi} \sum_{k \in S_N} \frac{1(y_k \leq t)}{\pi_{k|N}} \tag{1.7}$$

1.2.3 *Variance Computation*

The variance of the HT-estimator is

$$\begin{aligned} V\left(\hat{t}_{y\pi}\right) = & \sum_{k \in U_N} \pi_{k|N}(1 - \pi_{k|N}) \left(\frac{y_k}{\pi_{k|N}}\right)^2 \\ & + \sum_{k \neq l \in U_N} (\pi_{kl|N} - \pi_{k|N}\pi_{l|N}) \frac{y_k}{\pi_{k|N}} \frac{y_l}{\pi_{l|N}} \end{aligned} \tag{1.8}$$

In Eq. (1.8), the first term in the right-hand side is the variance we would obtain if the units in the population were selected independently, which is known as Poisson sampling (see Sect. 1.4.1).

A gain in efficiency can be obtained by using fixed-size sampling designs, which are such that only the subsets s in U_N of size n have positive selection probabilities $p_N(s)$. Many fixed-size sampling designs verify the so-called Sen–Yates–Grundy conditions:

$$\pi_{kl|N} \leq \pi_{k|N}\pi_{l|N} \text{ for any } k \neq l \in U_N \tag{1.9}$$

Under Eq. (1.9), the second-term in the right-hand side of (1.8) is non-positive for a non-negative variable of interest y_k, resulting in a variance reduction as compared to Poisson sampling.

For fixed-size sampling designs, the variance of the HT-estimator may be rewritten as

$$V\left(\hat{t}_{y\pi}\right) = \frac{1}{2} \sum_{k \neq l \in U_N} (\pi_{k|N}\pi_{l|N} - \pi_{kl|N}) \left(\frac{y_k}{\pi_{k|N}} - \frac{y_l}{\pi_{l|N}}\right)^2 \tag{1.10}$$

which is known as the Sen–Yates–Grundy (SYG) variance formula [53, 62].

Multinomial sampling (a.k.a. sampling with replacement) is an important benchmark for a sampling design $p_N(\cdot)$. It consists in n independent draws in the population with probabilities $p_{k|N} = n^{-1}\pi_{k|N}$ at each draw. Under multinomial sampling, t_{yN} is unbiasedly estimated by the Hansen–Hurwitz estimator

$$\hat{t}_{yHH} = \sum_{k \in U_N} W_{kN} \frac{y_k}{\pi_{k|N}} \tag{1.11}$$

with W_{kN} the number of selections for unit k, see [35]. The variance is

$$V\left(\hat{t}_{yHH}\right) = \sum_{k \in U_N} \pi_{k|N} \left(\frac{y_k}{\pi_{k|N}} - \frac{t_{yN}}{n}\right)^2 \tag{1.12}$$

The sampling design $p_N(\cdot)$ is more efficient than multinomial sampling if

$$V\left(\hat{t}_{y\pi}\right) \leq V\left(\hat{t}_{yHH}\right) \text{ for any variable of interest } y \tag{1.13}$$

Some sufficient conditions for Eq. (1.13) are listed in [29].

1.2.4 Variance Estimation

For any sampling design, the variance of the HT-estimator may be estimated by the HT-variance estimator

$$\hat{V}_{HT}\left(\hat{t}_{y\pi}\right) = \sum_{k \in S_N} (1 - \pi_{k|N}) \left(\frac{y_k}{\pi_{k|N}}\right)^2 + \sum_{k \neq l \in S_N} \frac{\pi_{kl|N} - \pi_{k|N}\pi_{l|N}}{\pi_{kl|N}} \frac{y_k}{\pi_{k|N}} \frac{y_l}{\pi_{l|N}} \tag{1.14}$$

For a fixed-size sampling design, we may alternatively use the SYG-variance estimator

$$\hat{V}_{SYG}\left(\hat{t}_{y\pi}\right) = \frac{1}{2} \sum_{k \neq l \in S_N} \frac{\pi_{k|N}\pi_{l|N} - \pi_{kl|N}}{\pi_{kl|N}} \left(\frac{y_k}{\pi_{k|N}} - \frac{y_l}{\pi_{l|N}}\right)^2 \tag{1.15}$$

These two variance estimators usually take different values.

Three properties related to the second-order inclusion probabilities are particularly useful for variance estimation. First and obviously, the $\pi_{kl|N}$'s need to be calculable. Second, the two variance estimators are design-unbiased if and only if all the $\pi_{kl|N}$'s are positive. Finally, for a fixed-size sampling design, the variance estimator $\hat{V}_{SYG}\left(\hat{t}_{y\pi}\right)$ is non-negative for any variable of interest if and only if the SYG conditions given in (1.9) are respected.

1.3 Simple Random Sampling

When the sampling design $p_N(\cdot)$ is defined, it is possible that no particular information is known on the population U_N, or that such information is meaningless for our target of inference. In such case, we have no other reasonable choice but to give to the units equal inclusion probabilities. In view of Eq. (1.3), this leads to

$$\pi_{k|N} = \frac{n}{N} \tag{1.16}$$

For example, equal inclusion probabilities are obtained under simple random sampling.

1.3.1 Definition

Simple random sampling without replacement (SI) is the fixed-size sampling design which gives to any subset $s \subset U_N$ of size n the same selection probability, namely

$$p(s) = \begin{cases} \frac{1}{\binom{N}{n}} & \text{if } Card(s) = n, \\ 0 & \text{otherwise} \end{cases} \tag{1.17}$$

Under SI, the HT-estimator simplifies as

$$\hat{t}_{y\pi} = N\bar{y}_N \text{ with } \bar{y}_N = \frac{1}{n}\sum_{k\in S_N} y_k \text{ the sample mean} \tag{1.18}$$

Also, the variance of the HT-estimator given in (1.8) simplifies as

$$V\left(\hat{t}_{y\pi}\right) = N^2\left(\frac{1}{n} - \frac{1}{N}\right)S_{yN}^2 \tag{1.19}$$

$$\text{with } S_{yN}^2 = \frac{1}{N-1}\sum_{k\in U_N}(y_k - \mu_{yN})^2 \text{ the population dispersion}$$

In the particular case of SI, both the HT-variance estimator and the SYG-variance estimator are identical, and equal to

$$\hat{V}_{HT}\left(\hat{t}_{y\pi}\right) = N^2\left(\frac{1}{n}-\frac{1}{N}\right)s_{yN}^2 \tag{1.20}$$

$$\text{with } s_{yN}^2 = \frac{1}{n-1}\sum_{k\in S_N}(y_k - \bar{y}_N)^2 \text{ the sample dispersion}$$

Among the large number of sampling algorithms proposed for SI, see Sect. 4.4 in [58] for a review, we consider three important implementations.

1.3.2 Draw by Draw Sampling Algorithm

The draw by draw procedure given in Algorithm 1 consists in successively drawing n units with equal probabilities, among the units remaining after the last draw. The procedure is computationally very intensive, since n readings of the file are necessary for sample selection. This is therefore not usable for sampling in large populations.

1.3.3 Selection–Rejection Method

The selection–rejection method [26] presented in Algorithm 2 requires one pass of the file only. We note $\mathscr{F}_{t-1}$ for the σ-field generated by the random variables up to step $t-1$: $\pi_{t|\mathscr{F}_{t-1}}$ is therefore the inclusion probability conditionally on selection steps $1,\ldots,t-1$, which in case of SI is simply the sampling rate among the remaining. Algorithm 2 is a fast implementation of SI, but improvements are possible, see, for example, [21] for a review.

We consider an illustration of the selection–rejection procedure on a small population of size $N=6$, presented in Table 1.1. We wish to select a sample of size $n=3$, and the inclusion probabilities are therefore $\pi_{k|N}=0.5$. Since $u_1 > \pi_{1|N}$, unit 1 is not selected in the sample. We have $j(1)=0$, and the conditional inclusion probability of units $k>1$ is therefore $\pi_{k|\mathscr{F}_1} = 3/5 = 0.60$. Since $u_2 \leq \pi_{2|\mathscr{F}_1}$, unit 2 is selected in the sample and $j(2)=1$. The conditional inclusion probability of

Algorithm 1 Draw by draw procedure for simple random sampling

- Initialize with $t=1$ and $U_N(1)=U_N$.
- For $t=1,\ldots,n$:
  - Select some unit (k, say) from $U_N(t)$ with equal probabilities.
  - Take $U_N(t+1) = U_N(t)\setminus\{k\}$.

Algorithm 2 Selection–rejection procedure for simple random sampling

- Initialize with $t = 0$. The number of selected units is $j(0) = 0$.
- For $t = 1, \ldots, N$:
  - Generate a random number u_t according to a uniform distribution.
  - If $u_t \leq \pi_{k|\mathscr{F}_{t-1}}$ the unit is selected, where

$$\pi_{k|\mathscr{F}_{t-1}} \equiv \frac{n - j(t-1)}{N - (t-1)}. \tag{1.21}$$

    In such case, $j(t) = j(t-1) + 1$.
  - Otherwise, $j(t) = j(t-1)$.

Table 1.1 Example of selection by the selection–rejection method

k	x_{0k}	u_k	$\pi_{k\|N}$	I_{kN}	$\pi_{k\|\mathscr{F}_1}$	I_{kN}	$\pi_{k\|\mathscr{F}_2}$	I_{kN}	$\pi_{k\|\mathscr{F}_3}$	I_{kN}	$\pi_{k\|\mathscr{F}_4}$	I_{kN}
1	30	0.87	0.50	0		0		0		0		0
2	10	0.09	0.50		0.60	1		1		1		1
3	20	0.68	0.50		0.60		0.50	0		0		0
4	10	0.18	0.50		0.60		0.50		0.67	1		1
5	20	0.28	0.50		0.60		0.50		0.67		0.50	1
6	30	0.59	0.50		0.60		0.50		0.67		0.50	0

units $k > 2$ is therefore $\pi_{k|\mathscr{F}_2} = 2/4 = 0.50$. The final sample is $S_N = \{2, 4, 5\}$, see Table 1.1.

1.3.4 *Reservoir Procedure*

The selection–rejection procedure requires to know the size N of the population to compute the inclusion probabilities. It is therefore not feasible when sampling in a data stream, when the population size is not fixed but grows over time. The reservoir procedure [45] presented in Algorithm 3 enables to select a SI sample, without knowing in advance the population size N. The principle is to maintain at any time t a reservoir S_t, which is in fact a SI sample of n units selected in $U_t = \{1, \ldots, t\}$. We consider in Sect. 1.4.5 a generalization to unequal probability sampling, called Chao's procedure [9].

We consider an illustration of the reservoir procedure on the population presented in Table 1.1. We initialize with $S_3 = \{1, 2, 3\}$. At Step 4, $u_4 \leq 3/4 = 0.75$, and unit 4 is therefore selected in the reservoir, while one unit of S_3 (2, say) is randomly removed. At Step 5, $u_5 \leq 3/5 = 0.60$, and unit 5 is therefore selected in the reservoir, while one unit of S_4 (4, say) is randomly removed. At Step 6, $u_6 > 3/6 = 0.50$, and unit 6 is therefore not selected. The final sample is $S_N = \{1, 3, 5\}$, see Table 1.2.

Algorithm 3 Reservoir procedure for simple random sampling

- Initialize with $t = n$, and with $S_n = \{1, \ldots, n\}$.
- For $t = n + 1, \ldots, N$, do:
  - Generate a random number u_t according to a uniform distribution.
  - If $u_t \leq n/t$, remove one unit (k, say) from S_{t-1} with equal probabilities. Take $S_t = S_{t-1} \cup \{t\} \setminus \{k\}$.
  - Otherwise, take $S_t = S_{t-1}$.

Table 1.2 Example of selection by the reservoir procedure

k	u_k	$\pi_{k\|N}$	I_{k3}	I_{k4}	I_{k5}	I_{k6}
1		0.50	1	1	1	1
2		0.50	1	0	0	0
3		0.50	1	1	1	1
4	0.18	0.50	0	1	0	0
5	0.28	0.50	0	0	1	1
6	0.59	0.50	0	0	0	0

1.3.5 Stratification

It is often possible to have access at the sampling stage to a q-vector x_k of auxiliary variables, known for any unit in the population. Rather than directly selecting a SI sample, we can use x_k to partition the population into groups. Sub-samples are then independently selected into each of these groups by SI. The groups are called strata, and the sampling design is called stratified simple random sampling (STSI).

We note $U_{Nh},\ h = 1, \ldots, H$ of size N_h the strata, in which SI samples S_{Nh} are independently selected. The HT-estimator may be rewritten as

$$\hat{t}_{y\pi} = \sum_{h=1}^{H} N_h \bar{y}_{Nh} \text{ with } \bar{y}_{Nh} = \frac{1}{n_h} \sum_{k \in S_{Nh}} y_k \text{ the sample mean in } S_{Nh} \qquad (1.22)$$

The variance of the HT-estimator may be rewritten as

$$V\left(\hat{t}_{y\pi}\right) = \sum_{h=1}^{H} N_h^2 \left(\frac{1}{n_h} - \frac{1}{N_h}\right) S_{yNh}^2 \qquad (1.23)$$

with S_{yNh}^2 the dispersion inside the stratum U_h. The comparison between Eqs. (1.19) and (1.23) shows that stratification reduces the variance by dropping the dispersion between strata, leaving the dispersion within strata only. STSI is therefore particularly efficient if the strata are homogeneous with respect to the variable of interest. A further gain in efficiency can be obtained by adequately allocating the global sample size into the strata [46].

Stratification is of routine use in surveys, and may result in large gains in efficiency. It is also useful if we wish to make estimations on sub-populations, and if these sub-populations may be used as strata. In this case, stratification makes it possible to control the sample size selected inside these sub-populations. Methods to define appropriate stratification of populations are discussed in [37] and [38].

1.4 Sampling with Unequal Probabilities

Suppose that a positive auxiliary variable x_{0k} is known for any unit in the population, and that this variable is positively correlated to y_k. For a fixed-size sampling design, it follows from Eq. (1.10) that the variance of the HT-estimator is small if $\pi_{k|N}$ is roughly proportional to y_k. Since y_k is unknown at the sampling stage, this suggests defining the inclusion probabilities proportionally to x_{0k}. This leads to probability proportional to size (π-ps) sampling, and the inclusion probabilities are

$$\pi_{k|N} = n \frac{x_{0k}}{\sum_{l \in U_N} x_{0l}} \tag{1.24}$$

Equation (1.24) may lead to inclusion probabilities greater than 1 for units with large values of x_{0k}. In such case, these inclusion probabilities are set to 1, and the other probabilities are recomputed until all of them are lower than 1.

For example, imagine that we wish to select in the population presented in Table 1.1 a sample of size $n = 5$, with probabilities proportional to x_{0k}. Applying Eq. (1.24) leads to

$$\pi_{1|N} = \pi_{6|N} = 1.25 \; \pi_{2|N} = \pi_{4|N} = 0.42 \; \pi_{3|N} = \pi_{5|N} = 0.83$$

We set $\pi_{1|N} = \pi_{6|N} = 1$, and recompute the other probabilities on the remaining population, i.e.,

$$\pi_{k|N} = (n-2) \frac{x_{0k}}{\sum_{l \in U \setminus \{1,6\}} x_{0l}} \text{ for } k \in U_N \setminus \{1, 6\}$$

The final inclusion probabilities are

$$\pi_{1|N} = \pi_{3|N} = \pi_{5|N} = \pi_{6|N} = 1, \; \pi_{2|N} = \pi_{4|N} = 0.5$$

In the rest of this section, we present some important sampling designs with unequal inclusion probabilities. A more comprehensive treatment of unequal probability sampling procedures may be found in [8] and [58].

Algorithm 4 Poisson sampling

- At step $t = 1, \dots, N$, do:
  - Generate a random number u_t according to a uniform distribution, independently of $u_1, \dots, u_{t-1}$.
  - The unit t is selected if $u_t \leq \pi_{t|N}$.

Table 1.3 Example of selection by Poisson sampling

k	x_{0k}	$\pi_{k\|N}$	u_k	I_{kN}
1	30	0.75	0.87	0
2	10	0.25	0.09	1
3	20	0.50	0.68	0
4	10	0.25	0.18	1
5	20	0.50	0.28	1
6	30	0.75	0.59	1

1.4.1 Poisson Sampling

Poisson sampling, which is described in Algorithm 4, is a very simple method for π-ps sampling. The units are selected through a series of N independent heads or tails experiment. In the particular case when all the units have the same inclusion probability $\pi_{k|N} = n/N$, the algorithm is called Bernoulli sampling.

For illustration, we consider the population in Table 1.1. The inclusion probabilities are defined proportionally to x_{0k}. All units are given independent random numbers u_k generated from a uniform distribution. Comparing them with the $\pi_{k|N}$'s leads to the sample $S_N = \{2, 4, 5, 6\}$, see Table 1.3.

The simplicity of Poisson sampling makes it attractive for coordination sampling, i.e., when we wish to select several samples with controlled overlap. Positive coordination targets a maximum overlap: this is useful in repeated surveys, where we are interested in comparisons over time. Negative coordination targets a minimum overlap: this is useful to minimize the response burden, and currently used in business surveys, for example. For a good overview of sample coordination methods, see, for example, [25, 49], and [31].

1.4.2 Rejective Sampling

Since the units are independently selected under Poisson sampling, the sample size is random. In the example given in Table 1.3, we targeted an average sample size of $n = 3$, but the size of the sample finally selected is 4. This random sample size results in a greater variability of estimators. To circumvent this problem, [34] introduced rejective sampling (a.k.a. conditional Poisson sampling) which is presented in Algorithm 5. It consists in repeatedly performing Poisson sampling,

Algorithm 5 Rejective sampling

- Let $\pi_{b|N} = (\pi_{b1|N}, \ldots, \pi_{bN|N})^\top$ denote a set of basic inclusion probabilities, with $\sum_{k \in U} \pi_{bk|N} = n$.
- Select a Poisson sample with inclusion probabilities $\pi_{b|N}$.
- We stop if the sample is of size n. Otherwise, we select another Poisson sample.

Table 1.4 Example of selection by rejective sampling

k	π_k	π_{bk}	u_{1k}	I_{kN}^1	u_{2k}	I_{kN}^2
1	0.75	0.71	0.87	0	0.92	0
2	0.25	0.29	0.09	1	0.28	1
3	0.50	0.50	0.68	0	0.85	0
4	0.25	0.29	0.18	1	0.21	1
5	0.50	0.50	0.28	1	0.69	0
6	0.75	0.71	0.59	1	0.16	1

with a basic set of inclusion probabilities $\pi_{bk|N}$, until the target sample size is attained. This is a fixed-size sampling design by construction. In the particular case of equal inclusion probabilities, rejective sampling is equivalent to SI.

Due to the rejection of some samples, the final inclusion probability $\pi_{k|N}$ differs from the basic inclusion probability $\pi_{bk|N}$. More precisely, denoting by S_{Np} the Poisson random sample and by S_{Nr} the rejective random sample, the final inclusion probabilities are

$$\pi_{k|N} \equiv Pr(k \in S_{Nr}) = Pr(k \in S_{Np}|Card(S_{Np}) = n) \tag{1.25}$$

[24] showed that for any prescribed inclusion probabilities $\pi_{k|N}$, associated basic probabilities $\pi_{bk|N}$ always exist. Some efficient algorithms to compute these probabilities are also available, see [15] and [18]. Rejective sampling has been widely considered in the sampling literature, see, for example, [44] for a review of variance approximations, and [6] for approximations of inclusion probabilities up to any order.

For illustration, we consider the population with inclusion probabilities $\pi_{k|N}$ given in Table 1.3. The associated basic inclusion probabilities $\pi_{bk|N}$ are given in Table 1.4. Using the first series of random numbers u_{1k}, we obtain a sample of size 4 which is therefore rejected. Using the second series of random numbers u_{2k}, we obtain the sample $S_N = \{2, 4, 6\}$ which is of appropriate size.

1.4.3 Systematic Sampling

Systematic sampling [43], which is presented in Algorithm 6, is a very popular method for π-ps sampling. It consists in ordering the units in the population, generating a random start to determine the first unit selected, and then making jumps

Algorithm 6 Systematic sampling

- Compute $V_0 = 0$ and $V_k = \sum_{l=1}^{k} \pi_{l|N}$ for any $k \in U_N$.
- Generate a random number u according to a uniform distribution.
- Select the units k such that for some integer $i \in \{1, \ldots, n\}$

$$V_{k-1} \leq u + (i-1) \leq V_k$$

Table 1.5 Example of selection by systematic sampling

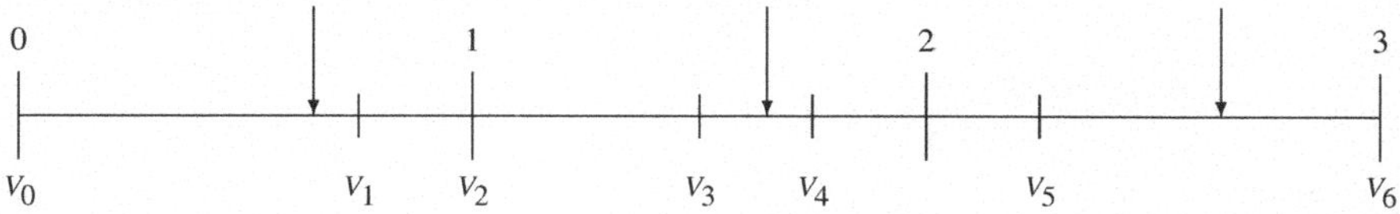

of size 1 to determine the other sampled units. This sampling design is very simple, but involves very few randomness since one random number only is sufficient for the whole sample selection, and its statistical properties are therefore limited.

For illustration, we consider the population in Table 1.3. We generate a random number $u = 0.65 \in [V_0, V_1[$, and unit 1 is therefore selected. By making jumps of length 1, we obtain the complete sample $S_N = \{1, 4, 6\}$ (Table 1.5).

There is a huge body of literature on systematic sampling, see, for example, [40] for a critical review. Systematic sampling enables selecting samples which are well-spread, which is in particular useful in spatial sampling. The popular generalized random tessellation stratified (GRTS) sampling [56] consists in defining an order on the spatial units by tessellation, and then applying systematic sampling; see also [60] and [3] for a review of spatial sampling methods.

1.4.4 Pivotal Sampling

Pivotal sampling [22], which is presented in Algorithm 7, consists in a succession of duels between units. At each step, we consider the two first units remaining in the population. If the sum of their probabilities is lower than 1, one of them gets the sum of their probabilities, while the other is discarded. If the sum of their probabilities is greater than 1, one of the units is selected, while the other one gets the residual probability. Pivotal sampling is also known in computer science as the Srinivasan sampling procedure, see [55].

An illustration of sample selection is given in Table 1.6. At the first step, units 1 and 2 fight, which results in selecting unit 1 and in discarding unit 2. At the second step, units 3 and 4 fight, and unit 3 gets the cumulative probability, while unit 4 is discarded. At the third step, units 3 and 5 fight, and unit 5 is selected, while unit 3 gets the residual probability $0.5 + 0.75 - 1 = 0.25$. At the fourth step, units 3

Algorithm 7 Pivotal sampling

- Initialize with $\pi(0) = \pi_N$.
- At step $t = 1, \ldots, T$:
  - Initialize with $\pi(t) = \pi(t-1)$.
  - Take $k < l$ the two first units in the population such that

$$\pi_k(t-1) \notin \{0, 1\} \text{ and } \pi_l(t-1) \notin \{0, 1\}$$

  - If $\pi_k(t-1) + \pi_l(t-1) \leq 1$, then do:

$$\{\pi_k(t), \pi_l(t)\} = \begin{cases} \{\pi_k(t-1) + \pi_l(t-1), 0\} \text{ with prob. } \frac{\pi_k(t-1)}{\pi_k(t-1)+\pi_l(t-1)} \\ \{0, \pi_k(t-1) + \pi_l(t-1)\} \text{ with prob. } \frac{\pi_l(t-1)}{\pi_k(t-1)+\pi_l(t-1)} \end{cases}$$

  - If $\pi_k(t-1) + \pi_l(t-1) > 1$, then do:

$$\{\pi_k(t), \pi_l(t)\} = \begin{cases} \{1, \pi_k(t-1) + \pi_l(t-1) - 1\} \text{ with prob. } \frac{1-\pi_l(t-1)}{2-\pi_k(t-1)-\pi_l(t-1)} \\ \{\pi_k(t-1) + \pi_l(t-1) - 1, 1\} \text{ with prob. } \frac{1-\pi_k(t-1)}{2-\pi_k(t-1)-\pi_l(t-1)} \end{cases}$$

- The algorithm stops at step T when all the components of $\pi(T)$ are 0 or 1. Take $I_N = \pi(T)$.

Table 1.6 Example of selection by pivotal sampling

k	π_k	$\pi_k(1)$	$\pi_k(2)$	$\pi_k(3)$	$\pi_k(4) = I_k$
1	0.75	1	1	1	1
2	0.25	0	0	0	0
3	0.50	0.50	0.75	0.25	0
4	0.25	0.25	0	0	0
5	0.50	0.50	0.50	1	1
6	0.75	0.75	0.75	0.75	1

and 6 fight, and unit 6 is selected, while unit 3 is discarded. The final sample is $S = \{1, 5, 6\}$.

Like systematic sampling, pivotal sampling enables to select samples which are well-spread over space. Since more randomization is introduced in the sample process, the method possesses better statistical properties, see Sect. 1.5. It has therefore been proposed as an alternative for spatial sampling, see, for example, [30] and [13].

Pivotal sampling is a particular case of the cube method [23], which enables to perform balanced sampling. If a q-vector x_k of quantitative variables is known for any unit $k \in U_N$ at the design stage, a sampling design is balanced on x_k if the random sample S_N is such that

$$\sum_{k \in U_N} I_{kN} \frac{x_k}{\pi_k} = \sum_{k \in U_N} x_k \tag{1.26}$$

with I_{kN} the sample membership indicator for unit k. This means that the HT-estimator of $t_{xN} = \sum_{k \in U_N} x_k$ exactly matches the known total, see [59] for a review on the cube method. Apart from the cube method, balanced samples may also be selected by rejective sampling. A sample is first selected by means of a basic sampling strategy. If Eq. (1.26) is approximately matched, we keep the sample. Otherwise, another candidate sample is selected by means of the basic sampling strategy [28, 34].

1.4.5 Chao's Procedure

Rejective sampling requires that the inclusion probabilities $\pi_{k|N}$ are computable from the start of the algorithm. Poisson sampling, systematic sampling, and pivotal sampling require that the probability $\pi_{k|N}$ is computable, at least when unit k is considered for selection. In view of Eq. (1.24), we therefore need to know at each step the total $t_{x_0 N} = \sum_{k \in U_N} x_{0k}$ of the auxiliary variable on the entire population.

This is feasible if we have a sampling frame of the units in the population. However, if we wish to sample in a data stream, the units arrive successively and $t_{x_0 N}$ is not known in advance. Chao's procedure [9], which is presented in Algorithm 8, enables to perform π-ps sampling without sampling frame. The presentation in Algorithm 8, due to [58], is somewhat simpler than the original algorithm.

We consider an illustration of Chao's procedure on the population presented in Table 1.1. We initialize with $S_3 = \{1, 2, 3\}$. At Step 4, $u_4 \leq \pi_{4|4} = 0.5$, and unit 4 is therefore selected in the reservoir, while one unit of S_3 is randomly removed with probabilities $p_{k|4}$ (in this particular case, unit 2 is removed with certainty). At

---

Algorithm 8 Chao's procedure

- Initialize with $t = n$, $\pi_{k|n} = 1$ for $k = 1, \ldots, n$, and $S_n = \{1, \ldots, n\}$.
- For $t = n + 1, \ldots, N$:
  - Compute the inclusion probabilities proportional to x_{0k} in the population U_t, namely:

$$\pi_{k|t} = n \frac{x_{0k}}{\sum_{k=1}^{t} x_{0l}} \tag{1.27}$$

    If some probabilities exceed 1, they are set to 1 and the other inclusion probabilities are recomputed until all the probabilities are lower than 1.
  - Generate a random number u_t according to a uniform distribution.
  - If $u_t \leq \pi_{t|t}$, remove one unit (k, say) from S_{t-1} with probabilities

$$p_{k|t} = \frac{1}{\pi_{t|t}} \left\{ 1 - \frac{\pi_{k|t}}{\pi_{k|t-1}} \right\} \text{ for } k \in S_{t-1}$$

    Take $S_t = S_{t-1} \cup \{t\} \setminus \{k\}$.
  - Otherwise, take $S_t = S_{t-1}$.

Table 1.7 Example of selection by Chao's procedure

k	x_{0k}	u_k	I_{k3}	$\pi_{k\|4}$	$p_{k\|4}$	I_{k4}	$\pi_{k\|5}$	$p_{k\|5}$	I_{k5}	$\pi_{k\|6}$	$p_{k\|6}$	I_{k6}
1	30		1	1	0	1	1	0	1	0.75	0.33	0
2	10		1	0.5	1	0	0.33		0	0.25		0
3	20		1	1	0	1	0.67	0.5	1	0.50		1
4	10	0.18	0	0.5		1	0.33	0.5	0	0.25	0.33	0
5	20	0.28	0				0.67		1	0.50	0.33	1
6	30	0.59	0							0.75		1

Step 5, $u_5 \le \pi_{5|5} = 0.67$, and unit 5 is therefore selected in the reservoir, while one unit of S_4 (4, say) is randomly removed with probabilities $p_{k|5}$. At Step 6, $u_6 \le \pi_{6|6} = 0.75$, and unit 6 is therefore selected while one unit of S_5 (1, say) is randomly removed with probabilities $p_{k|6}$. The final sample is $S_N = \{3, 5, 6\}$, see Table 1.7.

1.5 Statistical Properties of a Sampling Design

The choice of using some specific sampling design is based on both practical and theoretical matters. To produce consistent estimators with appropriate confidence intervals, it is desirable that the three following statistical properties hold: (a) the HT-estimator is weakly consistent for the true total; (b) the HT-estimator satisfies a central-limit theorem; (c) a consistent variance estimator is available for the HT-estimator.

In order to study the statistical properties of the sampling designs presented in this chapter, we consider the following assumptions:

- H1: There exists some constant $c_1, C_1 > 0$ such that for any $k \in U$:

$$c_1 \frac{n}{N} \le \pi_k \le C_1 \frac{n}{N} \tag{1.28}$$

- H2: There exists some constant C_2 such that:

$$\frac{1}{N} \sum_{k \in U} y_k^4 \le C_2 \tag{1.29}$$

- H3: There exists some constant $c_3 > 0$ such that:

$$c_3 \frac{N^2}{n} \le V(\hat{t}_{y\pi}) \tag{1.30}$$

Assumption (H1) states that the inclusion probabilities are not too variable, and do not depart much from those obtained when sampling with equal probabilities. This assumption is under the control of the survey sampler. It is assumed in (H2) that the variable of interest has a finite moment of order 4, and in (H3) that the variance of the HT-estimator is non-vanishing. These assumptions are fairly weak, although we may find situations under which they are not respected. For example, (H2) does not hold for heavily skewed populations where some units in the population have very large y_k's.

1.5.1 Weak Consistency of the HT-Estimator

The HT-estimator is weakly consistent for the total t_y if

$$N^{-1}\left(\hat{t}_{y\pi} - t_y\right) \rightarrow_{Pr} 0 \tag{1.31}$$

where $\rightarrow_{Pr}$ stands for the convergence in probability. Under assumptions (H1) and (H2), this property holds if the SYG conditions in (1.9) are respected. These conditions hold for SI, and for all the sampling designs presented in Sect. 1.4 except systematic sampling. The result is trivial for Poisson sampling, and has been proved by [15] for rejective sampling, [22] for pivotal sampling, and [9] for Chao's procedure. More generally, this property holds if the sampling design is negatively associated [7].

The weak consistency of the HT-estimator also holds under assumptions (H1) and (H2) if the sampling design is more efficient than multinomial sampling, see Eq. (1.13). This property holds true for SI. It does not generally hold for Poisson sampling and systematic sampling, but has been proved for the three other designs: see [51] for rejective sampling, [12] for pivotal sampling, and [54] for Chao's procedure.

1.5.2 Central-Limit Theorem

Several different methods have been used in survey sampling to prove the property (b), which states that the HT-estimator satisfies a central-limit theorem. For example, two different and very elegant proofs for SI are due to [32] and [33].

The simplest case occurs when the sampling design may be seen as a series of independent experiments, like for Poisson sampling where the units are selected independently, or for multinomial sampling where the draws are independent. In this case, the asymptotic normality follows from the Lyapunov central-limit theorem.

The other unequal probability sampling designs are more difficult to handle, due to the dependence in the selection of the units. Reference [32] introduced a coupling method to prove a CLT for simple random sampling, which he later extended in [34] to cover rejective sampling. The basic idea of the coupling method is to link the sampling design under study to another one where the units are selected independently, and such that the two sampling designs are close with respect to some distance function, see, for example, [57]. The coupling method has also been used by [48] for sequential Poisson sampling, and [11] and [14] for multistage sampling, see also [4].

Another technique consists in applying the weaker martingale CLT, see, for example, [36]. This is used by [47] for the Rao–Hartley–Cochran method, [48] for two-stage sampling designs, and [13] for pivotal sampling.

1.5.3 Consistency of a Variance Estimator

The property (c) that a weakly consistent variance estimator is available is somewhat difficult to prove, except for Poisson sampling for which it holds automatically from assumptions (H1) and (H2), and for SI for which it can be proved by tedious, but fairly straightforward computation, see exercise 2.21 in [2].

The second-order inclusion probabilities can be computed for all the sampling designs presented in Sect. 1.4, either by means of an explicit formula for systematic sampling [17] and pivotal sampling [10, 19] or by means of a recursive formula, see [18] for rejective sampling and [9] for Chao's procedure. However, it can be easily proved that many second-order inclusion probabilities are zero for both systematic sampling and pivotal sampling, and consequently unbiased variance estimators are not available. Numerous variance estimators have been proposed for systematic sampling, see, for example, [50, 61] or [27]. Variance estimators for pivotal sampling are considered in [12, 30] and [13].

Even for sampling designs with positive second-order inclusion probabilities, the consistency of the HT-variance estimator or the SYG-variance estimator requires in particular that the second-order inclusion probabilities are bounded below, which is difficult to prove. For rejective sampling, [34] proposed a variance approximation which does not require the second-order inclusion probabilities. This results in a simplified variance estimator, whose consistency is proved in [14].

1.6 Conclusion

Among the unequal probability sampling designs presented in this chapter, Poisson sampling, rejective sampling, and pivotal sampling possess good statistical properties. Poisson sampling has the disadvantage to lead to random sample size, and therefore to an inflated variance. Rejective sampling is presumably the best sampling method, presenting many favorable properties needed for statistical inference. Pivotal sampling also possesses good statistical properties, but no unbiased variance estimator is available. Anyway, conservative variance estimators are possible [13], and pivotal sampling leads to well-spread samples, and may be more efficient than rejective sampling when the units are ranked with respect to some variable which is related to the variable of interest.

Despite its great interest for sampling in data streams, the currently known statistical properties of Chao's procedure are limited. Reference [5] studies the validity of Hajek's variance estimator [34] for Chao's procedure, but under very restrictive conditions on the first-order inclusion probabilities. Establishing a CLT and the consistency of a variance estimator for Chao's procedure is both an important and challenging task.

References

1. Ardilly, P.: Les Techniques de Sondage. Editions Technip (2006)
2. Ardilly, P., Tillé, Y.: Exercices Corrigés de Méthodes de Sondage. Ellipses (2003)
3. Benedetti, R., Piersimoni, F., Postiglione, P., et al.: Sampling Spatial Units for Agricultural Surveys. Springer, Berlin (2015)
4. Berger, Y.G.: Rate of convergence to normal distribution for the Horvitz-Thompson estimator. J. Stat. Plan. Infer. **67**(2), 209–226 (1998)
5. Berger, Y.G.: Variance estimation with Chao's sampling scheme. J. Statist. Plann. Inference **127**(1–2), 253–277 (2005)
6. Boistard, H., Lopuhaä, H.P., Ruiz-Gazen, A., et al.: Approximation of rejective sampling inclusion probabilities and application to high order correlations. Electron. J. Stat. **6**, 1967–1983 (2012)
7. Brändén, P., Jonasson, J.: Negative dependence in sampling. Scand. J. Stat. **39**(4), 830–838 (2012)
8. Brewer, K.R., Hanif, M.: Sampling with Unequal Probabilities, vol. 15. Springer Science & Business Media, Berlin (2013)
9. Chao, M.: A general purpose unequal probability sampling plan. Biometrika **69**(3), 653–656 (1982)
10. Chauvet, G.: On a characterization of ordered pivotal sampling. Bernoulli **18**(4), 1320–1340 (2012)
11. Chauvet, G.: Coupling methods for multistage sampling. Ann. Statist. **43**(6), 2484–2506 (2015)
12. Chauvet, G.: A comparison of pivotal sampling and unequal probability sampling with replacement. Stat. Probab. Lett. **121**, 1–5 (2017)
13. Chauvet, G., Le Gleut, R.: Asymptotic properties of pivotal sampling with application to spatial sampling. Tech. rep. (2018)
14. Chauvet, G., Vallée, A.A.: Inference for two-stage sampling designs with application to a panel for urban policy. Preprint arXiv:1808.09758 (2018)
15. Chen, X.H., Dempster, A.P., Liu, J.S.: Weighted finite population sampling to maximize entropy. Biometrika **81**(3), 457–469 (1994)
16. Cochran, W.G.: Sampling Techniques, 4th edn. John Wiley & Sons, New York-London-Sydney (1977). Wiley Series in Probability and Mathematical Statistics
17. Connor, W.: An exact formula for the probability that two specified sampling units will occur in a sample drawn with unequal probabilities and without replacement. J. Am. Stat. Assoc. **61**(314), 384–390 (1966)
18. Deville, J.: Note sur l'algorithme de chen, dempster et liu. Document de travail non publié (2000)
19. Deville, J.C.: Une Nouvelle (encore une!) Méthode de Tirage à Probabilités Inégales. INSEE, Paris (1998)

20. Deville, J.C.: Variance estimation for complex statistics and estimators: linearization and residual techniques. Surv. Methodol. **25**(2), 193–204 (1999)
21. Deville, J., Grosbras, J., Roth, N.: Efficient sampling algorithms and balanced samples. In: Compstat, pp. 255–266. Springer, Berlin (1988)
22. Deville, J.C., Tillé, Y.: Unequal probability sampling without replacement through a splitting method. Biometrika **85**(1), 89–101 (1998)
23. Deville, J.C., Tillé, Y.: Efficient balanced sampling: the cube method. Biometrika **91**(4), 893–912 (2004)
24. Dupakova, J.: A note on rejective sampling. Contribution to Statistics (J. Hajek memorial volume). Academia Prague (1975)
25. Ernst, L.R.: The maximization and minimization of sample overlap problems: a half century of results. Bull. Int. Stat. Inst. **57**, 293–296 (1999)
26. Fan, C., Muller, M.E., Rezucha, I.: Development of sampling plans by using sequential (item by item) selection techniques and digital computers. J. Am. Stat. Assoc. **57**(298), 387–402 (1962)
27. Fewster, R.: Variance estimation for systematic designs in spatial surveys. Biometrics **67**(4), 1518–1531 (2011)
28. Fuller, W.A.: Some design properties of a rejective sampling procedure. Biometrika **96**(4), 933–944 (2009)
29. Gabler, S.: On unequal probability sampling: sufficient conditions for the superiority of sampling without replacement. Biometrika **71**(1), 171–175 (1984)
30. Grafström, A., Lundström, N.L., Schelin, L.: Spatially balanced sampling through the pivotal method. Biometrics **68**(2), 514–520 (2012)
31. Grafström, A., Matei, A.: Coordination of conditional Poisson samples. J. Off. Stat. **31**(4), 649–672 (2015)
32. Hájek, J.: Limiting distributions in simple random sampling from a finite population. Publ. Mat. Inst. Hung. Acad. Sci. **5**, 361–74 (1960)
33. Hájek, J.: Some extensions of the Wald-Wolfowitz-Noether theorem. Ann. Stat. **3**, 506–523 (1961)
34. Hájek, J.: Asymptotic theory of rejective sampling with varying probabilities from a finite population. Ann. Stat. **35**, 1491–1523 (1964)
35. Hansen, M.H., Hurwitz, W.N.: On the theory of sampling from finite populations. Ann. Stat. **14**, 333–362 (1943)
36. Helland, I.S.: Central limit theorems for martingales with discrete or continuous time. Scand. J. Stat. **9**, 79–94 (1982)
37. Hidiroglou, M.A., Kozak, M.: Stratification of skewed populations: a comparison of optimisation-based versus approximate methods. Int. Stat. Rev. **86**(1), 87–105 (2018)
38. Horgan, J.M.: Stratification of skewed populations: a review. Int. Stat. Rev. **74**(1), 67–76 (2006)
39. Horvitz, D.G., Thompson, D.J.: A generalization of sampling without replacement from a finite universe. J. Am. Stat. Assoc. **47**, 663–685 (1952)
40. Iachan, R.: Systematic sampling: a critical review. Int. Stat. Rev./Rev. Int. de Stat. **50**, 293–303 (1982)
41. Isaki, C.T., Fuller, W.A.: Survey design under the regression superpopulation model. J. Am. Stat. Assoc. **77**(377), 89–96 (1982)
42. Kiaer, A.: Observations et expériences concernant des dénombrements représentatifs. Bull. de l'Inst. Internat, de Stat. **9**, 176–183 (1896)
43. Madow, W.G., et al.: On the theory of systematic sampling, ii. Ann. Math. Stat. **20**(3), 333–354 (1949)
44. Matei, A., Tillé, Y.: Evaluation of variance approximations and estimators in maximum entropy sampling with unequal probability and fixed sample size. J. Off. Stat. **21**(4), 543 (2005)
45. McLeod, A.I., Bellhouse, D.R.: A convenient algorithm for drawing a simple random sample. J. R. Stat. Soc. Ser. C. Appl. Stat. **32**(2), 182–184 (1983)
46. Neyman, J.: On the two different aspects of the representative method: the method of stratified sampling and the method of purposive selection. J. R. Stat. Soc. **97**(4), 558–625 (1934)

47. Ohlsson, E.: Asymptotic normality of the Rao-Hartley-Cochran estimator: an application of the martingale CLT. Scand. J. Stat. **13**(1), 17–28 (1986)
48. Ohlsson, E.: Asymptotic normality for two-stage sampling from a finite population. Probab. Theory Relat. Fields **81**(3), 341–352 (1989)
49. Ohlsson, E.: Coordination of PPS samples over time. In: The 2nd International Conference on Establishment Surveys, pp. 255–264. Citeseer (2000)
50. Opsomer, J.D., Francisco-Fernández, M., Li, X.: Model-based non-parametric variance estimation for systematic sampling. Scand. J. Stat. **39**(3), 528–542 (2012)
51. Qualité, L.: A comparison of conditional Poisson sampling versus unequal probability sampling with replacement. J. Stat. Plan. Infer. **138**(5), 1428–1432 (2008)
52. Särndal, C.E., Swensson, B., Wretman, J.: Model Assisted Survey Sampling. Springer Series in Statistics. Springer, Berlin (1992)
53. Sen, A.R.: On the estimate of the variance in sampling with varying probabilities. J. Indian Soc. Agricultural Statist. **5**(1194), 127 (1953)
54. Sengupta, S.: On Chao's unequal probability sampling plan. Biometrika **76**(1), 192–196 (1989)
55. Srinivasan, A.: Distributions on level-sets with applications to approximation algorithms. In: Proceedings 2001 IEEE International Conference on Cluster Computing, pp. 588–597. IEEE, Piscataway (2001)
56. Stevens, D.L., Olsen, A.R.: Spatially balanced sampling of natural resources. J. Am. Stat. Assoc. **99**(465), 262–278 (2004)
57. Thorisson, H.: Coupling, stationarity, and regeneration. Probability and Its Applications. Springer-Verlag, New York (2000)
58. Tillé, Y.: Sampling Algorithms. Springer, New York (2006)
59. Tillé, Y.: Ten years of balanced sampling with the cube method: an appraisal. Surv. Methodol. **37**(2), 215–226 (2011)
60. Wang, J.F., Stein, A., Gao, B.B., Ge, Y.: A review of spatial sampling. Spatial Stat. **2**, 1–14 (2012)
61. Wolter, K.M.: An investigation of some estimators of variance for systematic sampling. J. Am. Stat. Assoc. **79**(388), 781–790 (1984)
62. Yates, F., Grundy, P.M.: Selection without replacement from within strata with probability proportional to size. J. R. Stat. Soc. Ser. B Methodol. **15**(2), 253–261 (1953)

Chapter 2
Core-Sets: Updated Survey

Dan Feldman

2.1 Motivation

Every day 2.5 exabytes of data (2.5×10^{18}) [69] are generated by cheap and numerous information-sensing mobile devices, remote sensing, software logs, cameras, microphones, RFID readers, and wireless sensor networks [58, 67, 87]. These require clustering algorithms that, unlike traditional algorithms, (a) learn unbounded streaming data that cannot fit into main memory, (b) run in parallel on distributed data among thousands of machines, (c) use low communication between the machines (d) apply real-time computations on the device, (e) handle privacy and security issues.

A common approach is to re-invent computer science for handling these new computational models, and develop new algorithms "from scratch" independently of existing solutions.

Data reduction/summarization advocates a different approach: given a large data set, reduce the data so that it takes up significantly less space in memory, but *provably* approximates the large data in a problem dependent sense. Running existing (possibly off-line, non-parallel, or inefficient) algorithms on the reduced (small) data would then produce a result that is provably close to the solution obtained from running on the complete (big) data. Such a compressed data set is sometimes called a core-set (or, coreset); see Fig. 2.1(left).

Figures by Ibrahim Jubran.

D. Feldman (✉)
Computer Science Department, University of Haifa, Haifa, Israel

F. Ros, S. Guillaume (eds.), *Sampling Techniques for Supervised or Unsupervised Tasks*, Unsupervised and Semi-Supervised Learning,
https://doi.org/10.1007/978-3-030-29349-9_2

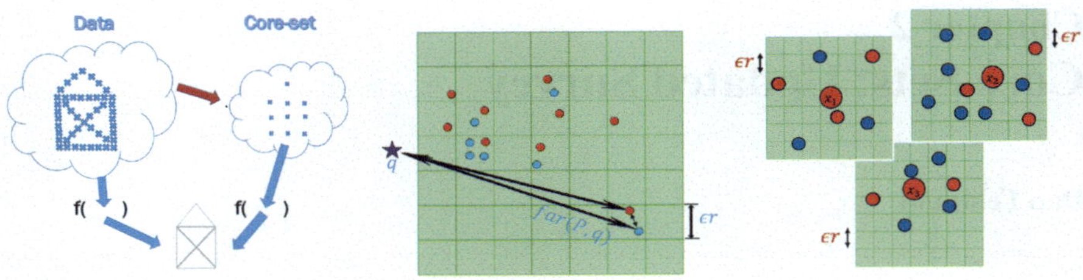

Fig. 2.1 **(left)** A function (algorithm) f gets a points set and outputs its clustering into linear segments. Applying f on the coreset of the data would ideally yield the same result faster using less resources, without changing f. **(middle)** An ε-coreset (in red) for 1-center queries $\text{far}(P, q) = \max_{p \in P} \|p - q\|_2$ of a set P of (blue) points and a query q, both on the plane, where $r = \min_{q' \in \mathbb{R}^2} \text{far}(P, q')$ is the minimum cost. **(right)** In k-center clustering, the query is replaced by a set Q of k centers, $\text{far}(P, Q) = \max_{p \in P} \min_{q \in Q} \|p - q\|_2$ and a grid is constructed around each of the optimal centers

Zen Coresets might be a better name for the coresets due to their flexible definition, based on the problem and related community: when we cannot find a coreset for a problem, we change the definition of a coreset. This is part of the reason why curious researchers and engineers find it hard to get into the coreset "cult."

The Coreset Paradigm shift is another source of confusion between readers, users, coreset developers, and especially reviewers: why your coreset construction algorithm assumes that it gets the optimal solution as input, if this is the main motivation for constructing it in the first place? Why you claim $O(n)$ construction time when the main theorem states $O(n^5)$? Where are the streaming and parallel coreset constructions (there are only off-line constructions)? Where is the analysis for the communication between the machines? What is wrong with uniform sampling as a coreset? Why do we need coresets, if gradient descent provides a sparse solution in linear time? What is the generalization error? Why there are experiments only on training (not testing) data?

The goal of this survey is to answer such questions by introducing the coreset paradigm together with the recent frameworks, as well as explaining their relations to other techniques.

Structure of This Paper The paper begins with high-level discussions and general definitions, continues to frameworks, and then to specific problems and solutions. Section 2.2 presents several inconsistent definitions of coresets that were suggested over the years, by focusing on their different properties, and provide a definition that captures most of them. Section 2.3 suggests example scenarios where coresets can be used or combined with existing solutions. Section 2.5 defines the type of problems that coreset may be applied on with some examples, while Sect. 2.4 defines specific types of coresets to solve such problems.

While there are dozens or hundreds of coreset constructions, Sect. 2.6 aims to present a framework called sup-sampling that can be used to solve many of them, by unifying existing frameworks. It usually reduces the coreset construction to the computation of the sup or max of corresponding functions, as well as their corresponding dimension (that is related to the notion of VC-dimension). To this end, Sect. 2.7 surveys techniques to bound sup and dimension for common family of problems.

Section 2.8 suggests future research and open problems.

2.2 What Is a Coreset?

The term coreset was coined in [3] and used for computing the smallest k balls that cover a set of input points, and then similar covering problems where coresets are called certificates (e.g. in [8]); see survey in [3]. Today there are many inconsistent coreset definitions and data reduction techniques adding to the motivation for writing this survey. We focus on coresets construction with provable guarantees for the trade-offs between size of coreset and approximation error. Other parameters such as construction or update time are usually derived from this trade-off; see Sect. 2.6.

Query Space is a tuple $(P, w, \mathcal{X}, f)$, where P is a (usually finite and of size $|P| = n$) input set that is called *points*, $\mathcal{X}$ is a (usually infinite) set of *queries* (models, shapes, classifiers, hypotheses), $w : P \to \mathbb{R}$ is a (usually positive) *weight* function that assigns importance to each point, and $f : P \times \mathcal{X} \to \mathbb{R}$ is a (usually non-negative) *pseudo-distance* function, or distance for short $f(p, x)$ from p to x; see [22, 40]. More generally it can be considered as a loss function. For max optimization problem it may be used as score (where higher is better).

Coreset in this review is a small data structure C that allows us to approximate the sum of weighted distances $\sum_{p\in P} w(p)f(p, x)$, its maximum distance $\max_{p\in P} w(p)f(p, x)$ which is related to *covering problems*, or in general, any function cost : $\mathbb{R}^n \to \mathbb{R}$ that maps the $n = |P|$ distances $(f(p, x))_{p\in P}$ to a (usually non-negative) total cost. The exact definitions for "small," "data structure," and "approximates" have been changed from paper to paper and we suggest a few of them in Sect. 2.4.

Example $P \subseteq \mathbb{R}^d$, $w \equiv 1$, $\mathcal{X} = \left\{X \subseteq \mathbb{R}^d \mid |X| = k\right\}$, and $f(p, X) = \min_{x\in X} \|p - x\|_2$ corresponds to k-means queries that aim to compute an *optimal query* set X^* of $|X^*| = k$ centers that minimize the sum of squared distances $\sum_{p\in P} f^2(p, X) = \left\|(f(p, X))_{p\in P}\right\|_2^2$ over $X \in \mathcal{X}$. More generally, $\mathbb{R}^d$ can be replaced by any set, and $\|p - x\|_2$ by any distance $f(p, x)$.In the k-median problem the cost is the non-squared distances $\sum_{p\in P} f(p, X) = \left\|(f(p, X))_{p\in P}\right\|_1$.

Composable Coresets [19, 62, 71] are important types of coresets that make them especially relevant for handling big or real-time data as explained in the next section. It means that a union of a pair of coresets $C_1 \cup C_2$ is a coreset for the underlying input $P_1 \cup P_2$, and that we can re-compute coreset C_3 for this coreset $C_1 \cup C_2$ recursively. In particular, strong and weak coresets as defined in Sect. 2.4.2 are composable.

Using merge-and-reduce trees [19] we can apply the coreset construction only on small subsets of the input independently (i.e., "embarrassingly parallel" [57]), merge them, and then reduce recursively. This implies that an efficient off-line coreset construction can be applied only to small subsets, to obtain linear coreset construction time via sub-linear memory; see Sect. 2.4.3.

2.3 Why Coresets?

In this section we give only a few of the many advantages of using coresets, or at least, composable coresets.

Answering Queries, such as SQL queries, to save either time, memory, or communication via the compressed small coreset C, including on the computation models below.

Optimization is the most common motivation for constructing coresets, where the goal is to compute an optimal query that minimizes the cost. Solving the optimization problem or its approximation on the small coreset yields an approximate solution of the original (big) dataset, sometimes after suitable post-processing. For example, the k-means problem is NP-hard when k is part of the input [76]. However, composable coresets of near-linear size in (k/ε) can be used to produce $1 \pm \varepsilon$ multiplicative factor approximation in $O(ndk)$ time, even for streaming distributed data in parallel [22, 52].

Boosting Existing Heuristics, i.e., algorithms with no provable guarantees is possible by running them on the small coreset. For example, we can run many more iterations or initial seeds on the coreset in the same time it takes for a single run on the original (big) data. In this sense we "improve the state-of-the-art using the state-of-the-art," and coreset is used as a bridge between theory and practical systems; see examples in [38, 51, 84].

"Magically" Turn Existing Off-Line Algorithms to (1) streaming algorithms, i.e., that use small memory and one pass over a possibly infinite input stream [19, 62], (2) parallel algorithms that use multiple threads (as in GPU), or more generally distributed data (network/cloud/swarm of robots or smartphones) via low or no communication between the machines [71], also simultaneously for unbounded stream of data [42] as in (1), and (3) dynamic (insertions and deletions of points) in small time, but using linear space [2].

This is done by maintaining a single coreset via these computational models during the night (or every few seconds). In the morning (or every few seconds) we apply the possibly inefficient existing algorithm in an off-line, non-parallel way from scratch on this coreset that represents all the data.

Constrained Optimization can be computed on a coreset that was constructed independently of these constraints. E.g. coreset for k-means queries can be used to compute the optimal query whose centers must be subset of the input, are close to some input points, or cannot be placed in forbidden areas, by solving the constrained problem on the coreset.

Model Selection and Features Reduction can be computed and evaluated on the coreset in a similar way. A particular useful property is that a coreset for, say, k centers is a coreset for $k' \leq k$ centers by definition: put k centers on top of k' centers. This allows us to handle regularization terms or costs that are common in machine learning such as to compute X that minimizes $\text{cost}_f(P, X) + |X|$ over every set X of $|X| \leq k$ centers [5, 12]. Same property occurs for other model parameters such as j-subspaces [52].

2.4 Coreset Types

Approximation Error for the majority of coresets aims to have a multiplicative $(1 + \varepsilon)$ approximation for the desired cost of each query $X \in \mathcal{X}$, or at least the optimal query. That is, an additive error of $\varepsilon\text{cost}_f(P, X)$. Grid coresets (See Sect. 2.4.3) usually introduce smaller error of $\varepsilon\text{cost}_f(P, X^*)$, where X^* is the optimal query. Weaker but sometimes smaller coresets use larger additive error that may be multiplicative under some assumption on the input (e.g., well clustered [33, 83], scaled data [36, 59]), by adding more parameters to the coreset's size that takes these parameters into account [49, 90], or lower bound on the cost for the allowed queries.

The Size of a coreset is usually by order of magnitude smaller than the size n of the input, about near-logarithmic or even independent of n. Rarely the reduction is on the dimension; see Low-Dimensional coresets in Sect. 2.4.1. The dependency on the approximation error $\varepsilon \in (0, 1)$ is usually polynomial in $1/\varepsilon$. Probabilistic constructions of coreset constructions are usually given a parameter $\delta \in (0, 1)$ that determines an upper bound on the probability of failure. The coreset size usually depends poly-logarithmically on $\log(1/\delta)$. This is strongly related to the size $O(\log(1/\delta)/\varepsilon^2)$ of a random sample that is needed to approximate the mean of a set of numbers in the interval $[0, 1]$, up to an ε error [68]. Rarely constructions the coreset size depends linearly on $1/\delta$, which is related to the fact that only δ fraction of the numbers are more than $1/\delta$ of the mean above [70]. First coresets were deterministic of size exponential in their (VC-)dimension d [3, 43], random constructions whose output is polynomial in d replaced them [26], and independency of d is sometimes possible using either weak coresets [42] or squared Euclidean distances [16, 52, 55].

2.4.1 Data Types

Any small coreset C may be used to answer queries efficiently. However, if the data type of the coreset is different from the original data type of P, we cannot use the same loss function f on C. In particular, we cannot directly run existing solvers for computing the optimal query on these coresets. More complicated and different data types may help to reduce the size of the coreset, but require the design of new (possibly non-efficient) optimization algorithms.

Weighted Subset of Input is a coreset $C \subseteq P$, where each point in C may be assigned a multiplicative scalar weight. The simplest case is where all the weights equal 1, and C is just a subset, which makes sense for covering problems [1, 2]. For approximating average of distances by few points, we can still avoid weights, e.g., by uniform sample/distribution [74]. For sum of weights, the weight is positive and intuitively tells us how many input points each coreset point represents [43, 62]. Most of the solvers support weighted input or can be easily changed to support it. However, negative weights or weights $u(p, x)$ that depend also on the query might imply non-convex optimization problem or other complications [40, 47, 50]; see Fig. 2.1(right).

Advantages of weighted subsets are (1) preserved sparsity of the input, (2) interpretability, (3) coreset may be used (heuristically) for other problems, (4) less numerical issues that occur when non-exact linear combination of points is used.

Weighted Subset of Input Space, where we assume that the input P is from a ground set or metric space, and the coreset is from the same ground set. E.g. for k-means, the coreset may be a subset of $\mathbb{R}^d$ but not a subset of P; e.g. [61, 86]. This is related to the notion of weak ε-net in computational geometry [79] (nothing to do with weak coresets in the next section).

Sketch Matrices imply that each point in the coreset is a linear combination of the input points. The input is represented as a matrix $P \in \mathbb{R}^{n\times d}$ where every point corresponds to a row, and the coreset construction is a "fat" matrix ($m \ll n$) $S \in \mathbb{R}^{m\times d}$ called *sketch matrix*. The coreset (called sketch) is then $C = SP \in \mathbb{R}^{m\times d}$. Sketch matrices are generalizations of previous coresets, where S is either a fat or sparse diagonal $n \times n$ matrix of the original row weights. As coresets, the term sketch is not consistent among papers. Sketch matrices may support the stronger turn-style model that allows deletion and changing single entries and not only insertion of records, in sub-linear space and without using trees. See surveys in [29, 85].

Low-Dimensional Coresets are coresets that instead of having small number of points, are contained in a low dimensional space in some sense. The classic examples are low-rank approximations (SVD/PCA) [30, 52, 77], JL-lemma (random projections) [21], and well conditioned matrices [32]. Sometimes such coresets are the first step before reducing the number of input points; see Sect. 2.7 for projections on the optimal query.

Generic Data Structures are usually the result of combining few types of the above coresets. These are the hardest to handle for optimization, but may be unavoidable for obtaining small coreset for some problems [40, 50, 86].

2.4.2 Query Sets

In Sect. 2.2 we defined coreset with respect to a set of queries that it approximates. The first two following types of coresets are much stronger than the third, as they are also composable.

Strong Coreset approximates *any* given query in the given query set. They are thus composable and it is usually easy to extract from them an approximation to the optimal query of the original input P. See examples in [3, 40, 85].

Weak Coreset is a coreset C that does not approximate all the queries. Instead, it approximates a subset of queries $\mathcal{X}(C)$. If this function $\mathcal{X}(C)$ is monotonic, i.e., $C' \subseteq C$ implies $\mathcal{X}(C) \subseteq \mathcal{X}(C')$, then the techniques for constructing ε-sample (see Sect. 2.6) can be applied [40]. Due to the smaller query space, the sample might be smaller by order of magnitude. This is since it depends on the generalized VC-dimension of the function $\mathcal{X}$ and not the VC-dimension of the set $\mathcal{X}(P)$; e.g. [40, 42, 45].

Example The VC-dimension for the range space of k balls in $\mathbb{R}^d$ is $O(dk \log k)$; [52]. However, a $(1 + \varepsilon)$-approximation to the optimal query (center) for the k-means/median/center problem is spanned by a convex combination of $k/\varepsilon^{O(1)}$ input points in P [55, 89]. By defining the function $\mathcal{X}$ that maps a coreset C to the union $\mathcal{X}(C)$ of all these centers, we obtain a generalized VC-dimension of $1/\varepsilon^{O(1)}$, i.e., independent of d for the query spaces of these problems. Generalizations hold for projective clustering $(j > 0)$, and suggestions on how to compute the optimal query of the restricted set $\mathcal{X}(C)$ can be found in [22, 40, 42, 45].

Sparse Solution (query) that approximates only the optimal query can be computed using convex optimization techniques such as Frank-Wolfe algorithm [13, 28]. An approximation solution, sparse or not, for a subset of the input points at hand, usually says nothing about the optimal solution of the complete data, or even after the insertion of a single new point. Hence, sparse solutions are not composable and thus do not support all the computation models from Sect. 2.3. Sometimes it is unavoidable: such coreset for 1-center has size $O(1/\varepsilon)$ for $\varepsilon \in (0, 1)$, while composable coreset or any coreset that handles streaming data must be of size exponential in d as was proven in [3]. Sometimes the sparse solution is the weight vector of the (strong) coreset itself [54].

2.4.3 Construction Types

Uniform Sample from the input is probably the most common "coreset." This is also a natural competitor of every other coreset. Unlike other constructions, uniform sampling takes sub-linear time in the input, which also explains why it misses small but important input points or far clusters, and does not provide $(1 \pm \varepsilon)$-multiplicative error as other coresets. Nevertheless, most of the coresets use uniform sampling after proper scaling by sup-weights which enables smaller coresets using the notion of ε-sample; see Sect. 2.6 and [31, 40, 74].

Importance Sampling aims to reduce the additive error of $\varepsilon n\text{cost}(P, x)$ to $\varepsilon\text{cost}(P, x)$, i.e., $(1 \pm \varepsilon)$-multiplicative error by replacing uniform sampling with non-uniform sample of the same size over the input space. The main technique is to re-weight each point by its sensitivity, or sup-weight respectively, as explained in Sect. 2.6. Then compute a "uniform" sample from the weighted set, where each input point is replaced by duplicated points according to its new weight, e.g., [22, 73].

Grids are based on discritization of the input space to small clusters, and then taking a representative from each cluster, weighted by the number of input points in its cell. These are the first coresets and were first used for covering problems [3, 7]. The additive error $\varepsilon cost(P, X^*) \leq \varepsilon cost(P, X)$ is usually smaller than modern coresets. However, the time and space is exponential in d due to the large number of cells, which is also the reason that deterministic constructions (that takes time exponential in d) are used. E.g. [7, 62]; see Fig. 2.1(middle/right).

Greedy Constructions are used for problems with specific properties to obtain smaller coresets, e.g., based on convex optimization. Here, in each iteration we adaptively pick the next best point to the coreset deterministically [54, 55] or based on importance sampling [9]. Such deterministic constructions may obtain coresets of size that cannot be obtained via random constructions; see Fig. 2.3(right), and Sect. 2.6 and [16, 17]

Such deterministic or adaptive constructions may be smaller by order of magnitudes compared to other constructions. For example, the query set of all the possible convex shapes (sets) has corresponding unbounded VC-dimension, but still has an ε-sample of finite size [25]; see also Lower Bounds in Sect. 2.6.

2.5 Problem Types

The Input Set P usually consists of finite number of $n = \sum_{p \in P} w(p)$ points in some metric or Euclidean space. Some coresets have size independent of n, and may be applied on a set P that consists of infinite number of points, e.g., an analog and continuous signal or image [45, 49, 61, 86]. Every point may be assigned a positive multiplicative weight, which usually does not make the problem easier or harder. This is necessary when computing coreset for coresets as in the case of composable coresets.

The Fitting Error f is usually a Log-Lipschitz function of some distance function and thus satisfies the weak triangle inequality in some sense, i.e., $f(p, x) - f(p', x)$ is bounded by the distance from p to p' for every $p, p' \in P$ and $x \in X$, up to a small multiplicative factor. This includes a distance to the power of a constant $z \geq 0$ or M-estimators that ignore large distances and thus handle outliers. When the query X is a set of centers as in k-means, or a shape as in low-rank approximations, or k-line means, we usually define $f(p, X) = \min_{x \in X} f(p, x)$; see [22, 41].

Projective Clustering is one of the fundamental type of problems that is suitable for coresets [4, 6, 34, 53, 64]. Here, each item in the set of queries $\mathcal{X}$ is a set of k affine j-dimensional subspaces. Special cases include k-means/median/center ($j = 0$) in a metric or pseudo-metric spaces (discrete versions) [23, 24], low-rank approximation (PCA) where $k = 1$ [24, 35], or well conditioned basis [32] for different distance functions. Many other problems can be reduced to these kinds of problems via linearizations in high-dimensional space. Projective clustering is a special case of dictionary learning as explained in [39].

Supervised Learning is done recently using coresets, where each input point (p, y) includes a discrete or continuous label. E.g. for learning kernels, $y \in \{-1, 1\}$, and the distance function is $f((p, y), x) = y \cdot \phi(p \cdot x)$ for some kernel function $\phi : \mathbb{R} \to [0, \infty)$. Unlike other optimization techniques, if the coreset is composable we may compute a coreset for each class y independently and then return the union of coresets as the final coreset [72, 75, 90].

Generalization Error is less relevant for most of the coresets. Unlike in machine learning and more like in computer science and computational geometry, there is usually no assumption that the data is a set of i.i.d samples from some known or unknown distribution. The question of what problem to solve, using which model and how to avoid overfilling is related to techniques such as maximum-likelihood or empirical risk-minimization. Exceptions include model selection in Sect. 2.3.

2.6 Generic Coreset Construction

Many of the existing and especially old coreset constructions can be simplified, improved, and have better guarantees using modern and retrospective analysis. In this section we try to give a generic algorithm that can be used to generate many of these existing coresets, which continues and improves such tries from [22, 40].

Reduction to Off-Line Construction on a Pair of Coresets is possible using composable coresets (strong or weak) construction which support the streaming model as explained in Sect. 2.2. Using the merge-and-reduce tree (as explained in Sect. 2.2), we can assume that we are given a set of $m \ll n$ points, which is a union of two other coresets, and the goal is to reduce it by half to a coreset of size $m/2$, for the smallest possible value of m. We then partition the (possibly infinite) input stream into consecutive subsets of size m. Each inner node of the merge-reduce tree then gets the union of coresets in its pair of children, and reduce them from $2m$ to m. In this way every level has at most one coreset in memory and the overall coreset is the union of coresets over the levels of the tree. This assumption usually adds extra polynomial factors in $\log(n)$, since ε is replaced with $\varepsilon/(\log n)$ due to the increasing error over the levels of the tree. The final coreset can still be small by constructing it once from the union of coresets above with ε, and not $\varepsilon/\log(n)$.

ε-Sample Many, if not most, of the existing coresets can be defined formally as an ε-sample, which is a generalization of the definition of ε-nets in computational geometry for binary functions [37, 66, 74]. Given a query space $(P, w, \mathcal{X}, f)$ and an error (usually constant) $\varepsilon \in (0, 1)$, an ε-sample is a query space $(C, u, \mathcal{X}, f)$ where $C \subseteq P$, such that $\text{cost}((w(p)f(p,x))_{p\in p}) = \sum_{p\in P} w(p)f(p,x)$ is approximated by $\text{cost}(u(q)(f(q,x))_{q\in C})$ for every query $x \in \mathcal{X}$, up to an additive error of ε, i.e.,

$$\left| \sum_{p\in P} w(p)f(p,x) - \sum_{q\in C} u(p)f(q,x) \right| \leq \varepsilon. \tag{2.1}$$

Hence, C is a weighted subset strong coreset, and is also a composable coreset as defined in Sect. 2.2.

In most papers related to coresets we were actually interested in $(1+\varepsilon)$-multiplicative error, and not an additive error of ε, e.g., since the problem hardness is invariant to scaling. In this case, the common $(1 \pm \varepsilon)$-multiplicative approximation error for a function $\sum_{p\in P} w(p)g(p,x)$ can be obtained by defining $f(p,x) = g(p,x)/\sum_{q\in P} w(q)g(q,x)$. Of course, an additive ε-approximation for f as in 2.1 is a multiplicative $(1+\varepsilon)$-approximation for the original function g.

Smaller bound on the error as in Grid coresets may be obtained by defining $f(p,x) = g(p,x)/\sum_{q\in P} w(q)g(q,x^*)$, where x^* is an optimal query. If this is hard or impossible, larger (no-longer multiplicative) error of $\Delta(P,x) \geq 0$ can be obtained by defining $f(p,x) = g(p,x)/\Delta(P,x)$.

Simple ε-Sample If the input weights are non-negative, i.e., $w : P \to [0, \infty)$, then PAC-learning [20, 65, 74], which generalizes Hoeffding's inequality [68], proves that a simple i.i.d uniform sampling (if $w \equiv 1/n$, or proportional to w in general) yields an ε-sample with probability at least $1-\delta$. The size of the sample depends polynomially on $\max_{p\in P, x\in\mathcal{X}} |f(p,x)|$, the sum of weights $\sum_{p\in P} w(p)$, $1/\varepsilon$, $\log(1/\delta)$, and a measure of complexity for a query space that is related to generalized VC-dimension [22, 40]. Formally, given a pair $(F, ranges)$ where F is a set, and $ranges$ is a set of subsets from F, the VC-dimension of $(F, ranges)$ is the size $|G|$ of the largest subset $G \subseteq F$ such that

$$|\{G \cap range \mid range \in ranges\}| = 2^{|G|}.$$

Intuitively, this is usually (but not always) the number of free parameters that are needed to define a single range in the set, i.e., query in the query set. For example, in k-means the corresponding range space is the set of k-balls in $\mathbb{R}^d$, which has VC-dimension $O(kd)$. Indeed each ball in $\mathbb{R}^d$ can be defined by its center and radius ($d+1$ parameters), and thus k balls require $k(d+1) = O(kd)$ such parameters.

While uniform sample is the most common and generic way to compute an ε-sample, generic deterministic constructions (usually of size exponential in d) were suggested in [78], and obtain the same or better bounds of most of the Grid coresets, after proper scaling by the sup-weights below. See Sect. 2.4.3 and Fig. 2.1(right).

Sup-Weights were suggested in [40] to reduce the size or the error of the sample above by a factor of n. We can see that it is possible to replace the weight function w by any function $m : P \to [0, \infty)$ and $g(p, x) = w(p)f(p, x)/m(p)$, and obtain the same desired cost $\sum_{p \in P} w(p)f(p, x) = \sum_{p \in P} m(p)g(p, x)$. The sample size above now depends on $\max_{p,x} |g(p, x)|$, and the new total weight $\sum_{p \in P} m(p)$. To minimize them, we assign to each point $p \in P$ a weight $m(p) = w(p) \max_{x \in \mathcal{X}} |f(p, x)|$ (more generally, its supremum). That is, we minimize the total weight under the constraint

$$\max_{p,x} |g(p, x)| = \max_{p,x} \frac{w(p)|f(p, x)|}{m(p)} = 1.$$

Computing weaker bounds $m'(p) \geq m(p)$, say, up to a multiplicative constant factor for every point $p \in P$, would increase the total weight and size of resulting coreset by the same factor.

This new optimization problem is independent of ε or one of the many algorithms to construct ε-sample (randomly or deterministically) on the new re-weighted set. Hence, it simplifies the analysis and results of many previous papers.

Sensitivity is a special case of this reweighting, where we wish to obtain multiplicative $1 \pm \varepsilon$ approximation for a non-negative function f, i.e., $f(p, x) = g(p, x)/\sum_{q \in P} w(q)g(q, x)$ as explained in the beginning of this section. Here,

$$m(p) = \max_x w(p)f(p, x) = \max_x \frac{w(p)g(p, x)}{\sum_{q \in P} w(q)g(q, x)}.$$

See e.g. [22, 26, 48, 73].

Finding small sup-weights or sensitivities for different cost functions is the main challenge of many current coreset papers. Sometimes it may be harder than the original optimization problem of finding x^*. In fact, the problems are usually related as explained below.

The Chicken-and-Egg problem stems from the fact that the optimal query x^* is usually required to bound sensitivity or sup-weights; see Fig. 2.1(right) and Sect. 2.7. However, this is usually the main motivation for constructing the coreset in the first place. There are a few lee-ways: first, it can be assumed that the size of the input is the size of a pair of coresets $m \ll n$ as explained in the beginning of this section. So inefficient algorithms (say, polynomial in their input size m), would still yield linear construction time in n. However, for NP-hard problems such as k-means clustering, the running time would still be exponential in k. For other problems (such

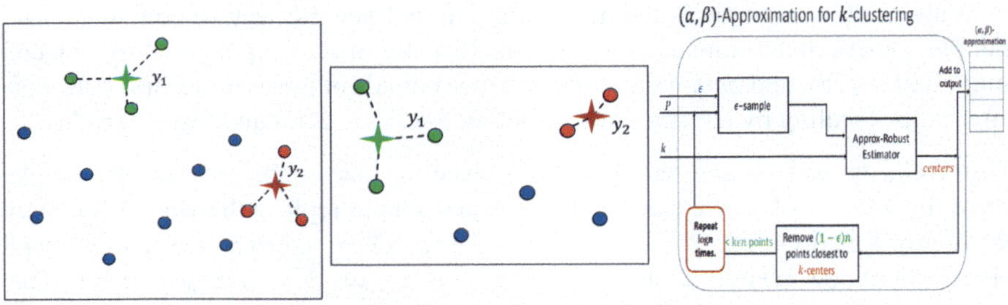

Fig. 2.2 **(left)** A robust approximation to the optimal clustering (the stars) is computed on a small εn-sample. About half of the closest points in the full set are then removed. **(middle)** The process continues recursively on the remaining half and new robust approximation (the new stars) is produced, until there are no points are left. **(right)** The output (α, β)-approximation is the union of robust estimators during the $O(\log n)$ iterations

as general projective clustering) even inefficient optimization algorithms might not be known.

Bicriteria or (α, β)-Approximation for the optimal solution can be used in such cases, which is a set $B \subseteq \mathcal{X}$ of $\beta \geq 1$ queries whose fitting cost $\sum_{p \in P} \min_{x \in B} f(p, x)$ is larger by a factor of at most $\alpha \geq 1$ compared to the cost of the optimal solution; see Fig. 2.2. The bound on the total sup-weights or sensitivities usually depends polynomially on α and β so near-logarithmic bounds for α, β are reasonable.

Generic Algorithm for (α, β)-Approximation is suggested in [40] following [44]. First we compute an $(\alpha, \beta, 1/4)$-approximation $X \subseteq \mathcal{X}$ to the *robust optimal* query as explained below, then we compute for each point $p \in P$ its distance $\min_{x \in X} f(p, x)$ to its closest center in X, and remove half of the closest points (or weights, in general). We continue recursively on the remaining $n/2$ input points (or total weight) until there are no points left; see Fig. 2.2. The running time is only linear in the number of input points due to the geometric sequence of points in each iteration.

Unlike techniques such as RANSAC [27], while the first iteration uses random sampling, the overall algorithm is adaptive: in each layer we remove the "main stream/cluster" of the data. "Hidden isolated" clusters may not be caught by uniform sampling in the first iteration but would be left and be discovered during the last iterations.

Robust Optimal query $x \in \mathcal{X}$ minimizes the sum of distances to its closest (say, half) input points $P_{x,1/2}$, i.e., ignoring a constant fraction (1/2) of outliers. Computing such a query is usually much harder than computing the optimal query x^*. However, computing an approximation $\tilde{x}$ that serves only quarter of the points, such that $\sum_{p \in P_{\tilde{x},1/4}} f(p, \tilde{x}) \leq \sum_{p \in P_{x,1/2}} f(p, x)$ is much easier, since ε-sample (in particular, uniform sample) is a coreset for this problem [40]. An $(\alpha, \beta, 1/4)$

approximation allows multiplicative factor α of error to the last cost, and using $|X| = \beta$ centers.

For example, for problems such as k-means/median/center, an ε-sample S for k balls, i.e., uniform sample of size $|S| = O(k)$ is also the desired $(1, |S|, 1/4)$-robust estimator. Generic algorithm for computing such robust estimator can be found in [40]. In practice, it may be computed via heuristics such as EM-estimators, but without provable bounds [48, 80].

Bootstrapping is used to reduce the overall size of the ε-coreset after computing an ε-sample that is based on sup-weights that are in turn based on an (α, β)-approximation. These total sup-weights or sensitivities usually depend polynomially on α and β, and so does the final coreset. To remove these factors we compute the coreset (off-line, on each subset of the merge-and-reduce tree) using, say, $\varepsilon' = 1/2$ to obtain a small coreset that can be used to compute a constant factor approximation to the optimal query ($\alpha' = \beta' = O(1)$). We then compute the coreset for the desired $\varepsilon \in (0, 1)$ using this (α', β')-approximation instead of the previous (α, β)-approximation to obtain coreset that depends on $\alpha' = \beta' = O(1)$. See e.g. [63].

In this sense, the framework in this section can be seen as a series of improved approximations, from initial (εn)-samples that are based on uniform samples, to sup-weights sampling that are based on a rough (α, β)-approximation that is in turn based on robust estimators in each iteration, which are eventually replaced by (α', β')-approximation and smaller total sup-weights/sensitivity. This process is then applied on each small subset (node) of the merge-reduce tree.

Lower Bounds Since there is no exact definition for a coreset, the most general definition is "any data structure that can answer every query in the query set." Here, the computation issues from Sect. 2.4.1 are ignored. The size of the coreset can be measured by the total number of bits it takes to store it in memory. A lower bound under this assumption is usually proved via communication protocols where Alice tries to answer a query based on an input data that Bob has, using minimum communication between them [81, 82].

Other lower bounds are known for specific problems and coreset types, such as for the size of weighted subset coresets [36, 59, 60], dependency on the stretch ratio of the input [92], or order of input points [11]. For random constructions, lower bounds may be computed using e.g. the coupon collector [22], and smaller deterministic versions are possible via [17].

2.7 Bounding Total Sup-Weights and Dimension

In this section we suggest generic existing technique to bound sup-weights and dimension of query space.

2.7.1 Bounding Sup-Weight

Most of the techniques for bounding total sensitivity or sup-weights in general are based on a given (α, β)-approximation for the optimal query of the corresponding query space, as explained in the previous section. For simplicity, we assume in this section that we are given the optimal solution ($\alpha = \beta = 1$). Otherwise we use the bootstrapping technique above.

Grids First coresets discretized/clustered the space into a grid of cells around the optimal solution, or around each center in the optimal solution, as explained in Sect. 2.4.2 and Fig. 2.1(right), such that every input point p in a grid cell $\square$ has the same distance to every query up to an additive factor of $\varepsilon f(p, x^*)$. Hence, the sup-weight is

$$\frac{f(p,x)}{f(P,x^*)} \leq \frac{f(p,x)}{f(P \cap \square, x^*)} \leq \frac{1}{|P \cap \square|}.$$

The sum of the last term over every point $p \in P \cap \square$ in the square is 1, so the total sup-weights is the number of cells in the grid, which is usually exponential in d. This is why deterministic ε-sample constructions, whose time is exponential in d, are used in these coreset constructions. The approximation error for $f(P, x)$ is $\varepsilon f(P, x^*) \leq \varepsilon f(P, x)$ as explained in Sect. 2.6.

Projection on the Optimal Query is an inductive technique to bound sensitivity, which is also used for solving related optimization problems. The idea is to compute the optimal solution or its approximation and to project the input on this subset (usually a shape or set of shapes). Assuming that we are using a metric space, by the weak triangle inequality the distance from the projected set and the original set to any query is bounded by the optimum cost, which is less than the query's cost. We then add this projected input points P' to the coreset, and compute coreset for the difference $f(P, x) - f(P', x) \leq f(P, P')$, by bounding the sup-weight of $f(p, x) - f(p', x) \leq f(p, p')$. Here we assume that $f(P, P')$ and $f(p, p')$ are well defined. The resulting coreset is the union of P' with a weighted subset of pairs that can be replaced by two points: p with positive weight and p' with a negative weight, as explained in [47, 50]. For problems such as k-means we have that P' consists of only k points, and the negative weights can actually be removed [22, 40]. For other problems we need to compute a coreset for P' of size $|P| = n$ but whose dimension is smaller. See following techniques.

Bounding Sensitivity directly using the previous technique is possible since the sensitivity of a point is bounded by $f(p, x^*)/f(P, x^*)$ plus the sensitivity of its projected point p' above with respect to the set P', as proven in [93]. The coreset is usually larger compared to previous technique (e.g., for k-means the total sensitivity is k and not 1). However, the resulting coreset is a weighted subset, with only positive weights.

The cUTE Decomposition is used to bound sensitivity of higher-dimensional centers than points, such as in projective clustering where $j \geq 1$. Using the above techniques we can reduce the dimension of P to the dimension j of the optimal query by replacing it with P'. It was proved in [43, 47] that in this case every m-dimensional affine subspace X can be replaced by an affine $(j-1)$-dimensional subspace X' and a constant $c > 0$ such that the distance from every point in P' to X is the same as the distance to X' multiplied by a constant (weight) c. See Fig. 2.2(left) for the case $j = 1$. This is a special case for the claim that there is a factorization $A = cU^T E$ for every $A \in \mathbb{R}^{m \times j}$ and $E \in \mathbb{R}^{d \times j}$, such that $U^T U = I$ and $c \geq 0$.

The reduction is from a set P in $\mathbb{R}^d$ and j-dimensional queries, to a set P' in $\mathbb{R}^j$ with j-dimensional queries. We can then apply this reduction recursively $j-1$ times till the centers (queries) are points. Unfortunately, careful analysis shows that the final total sensitivity is exponential in j [47].

Weighted Centers (multiplicative weight $w'(x)$ for every query x) occur: (1) when clustering to $k \geq 2$ (multiple) subspaces, which reduce to handling weighted $(j-1)$-subspaces as explained above, and (2) for clustering weighted facilities (centers), e.g., when the travelling costs to a center at the sea/air and a center of the same distance on the ground are not the same, and (3) handling k-clustering with m outliers (centers whose weight is infinity), or M-estimators $f'(p, x) = \min\{f(p, x), c)\}$ for some constant threshold $c > 0$.

Bounds on the sensitivities for these cases are usually exponential in the number of centers and also depend on $\log(n)$; See Sect. 2.4.3.

Reduction of cost(·) from $\|\cdot\|_1$ to $\|\cdot\|_\infty$ Suppose that we have a subset coreset construction for our query space, $\varepsilon = 1/2$, and any input P of n points, where $\text{cost}(\cdot) = \|\cdot\|_\infty$, i.e., covering queries. If the output coreset consists of at most m points, then it was proven in [52] based on a simple generalization of [92] (for projective clustering), that the total sensitivity is bounded by $O(m \log n)$.

Beyond Sup-Weights The max-sampling framework is very generic but for specific problems we might get better coreset constructions. This includes pre-processing techniques such as dimension reduction before computing the coreset [52]. The sparsity of the input is loss in this case, so recursive partitioning of the data can be used instead [16]. Convex optimization techniques such as the Frank-Wolfe algorithm [14, 28] produces a sparse solution for a given function which are not composable, as explained in Sect. 2.4.2. However, a coreset construction may be formulated as such a convex optimization problem where we wish to compute a sparse distribution over the input (the set of coreset weights) that satisfies some requirements. See for example deterministic small coresets for 1-mean and low-rank approximation [54, 55].

2.7.2 Bounding Queries Dimension

The dimension of a query space is the VC-dimension of the corresponding range space of the ε-sample as explained in Sect. 2.6. There are many techniques for bounding the VC-dimension of such a range space; see e.g. [10]. The range space is the same for the same pseudo-distance function to any power of $z \geq 1$. For example, k-median/median/center queries have the same corresponding range space: the sets of k balls in $\mathbb{R}^d$. To bound the VC-dimension it is usually useful to consider the squared distances, which are polynomial functions in case of points in $\mathbb{R}^d$, as in the case of projective clustering. The VC-dimension for such n polynomial functions is their degree d [52]. More generally, if we can answer a query in $O(d)$ arithmetic operations and exponential functions then the corresponding dimension of the query space is also polynomial in d.

Weak Composable Corests maintains the approximated optimal solution using smaller sample, by replacing the VC-dimension by generalized VC-dimension that may be smaller by order of magnitudes, or even independent, on parameters such as dimension d; see Sect. 2.4.2.

Homomorphic Query Space is a query space whose dimension is smaller, but sufficient to answer every query of a query space of a larger dimension. For example, squared distances to any shape that is contained in a k-subspace can be approximated up to factor of $1 \pm \varepsilon$ by another rotated shape that is contained in a fixed (k/ε)-dimensional subspace [52].

Deterministic Constructions for ε-sample might be smaller than the non-tight bound on the size that is obtained from uniform random sampling or other techniques. See Fig. 2.3(right) and Sect. 2.4.3.

Fig. 2.3 **(left)** For every pair of lines ℓ (in blue) and ℓ' (in red) there is a center c (in green) and $w \geq 0$ such that the distance from every $p \in \ell$ to ℓ' is the same as its distance to c multiplied by w. **(middle)** An arbitrary input point c_1 is the first center and coreset point. Its farthest input point p_2 is the second coreset point, and c_2 is the closest point between them to the origin. **(right)** The next coreset point p_3 is the farthest input point from c_2, and c_3 is the new closest point to the origin. After $i = 1/\varepsilon$ iterations we have $\text{error}_i = \|c_i\| \leq \varepsilon$

2.8 Future Research

Another direction is to apply coresets for problems beyond data reductions, as the following two examples.

Private Coresets In the recent years, we started to solve theoretical and practical open problems using coresets in fields that seemingly have nothing to do with data reduction. For example, in [46, 56] *private coresets* that preserve differential privacy were suggested to enable answer unbounded number of queries with no leakage of privacy for a specific user. It was also proved there that a small coreset for *any* problem implies a private coreset that introduces a small additive noise, for the corresponding query space. However, the proof is not constructive. Having a constructive proof will turn the existing numerous coreset constructions into private coresets for many open problems in machine learning. Such a contribution is especially important since lack of practical results for machine learning is one of the critique on differential privacy [15]. A more humble but important result is to suggest private coresets for specific problems, based on their corresponding non-private versions as in [46, 56].

Deterministic Coresets Using the sup-sampling approach in this survey, the problem of computing coresets can be reduced to the problem of computing ε-samples, after proper weighting by the sup or sensitivity of the function at hand. Computing such ε-samples deterministically of size polynomial in the VC-dimension is an open problem even for the case where the query is the set of half-spaces in $\mathbb{R}^d$. Promising direction is suggested in [91].

Deep Learning Coresets for deep learning are natural since deep learning usually applied on very big data sets, in parallel, and the training time is long. One goal may be to reduce the training data for shorter training time, and another goal may be to sparsity (compress) the network itself for faster classification. Since the functions in deep learning are much more complicated than the functions in this survey, a natural approach is to compute coreset for each neuron and then train the network neuron-by-neuron as in [90]. Similarly, we can compress the network neuron-by-neuron via sensitivities of edges as was suggested in [18, 88]. Since no coresets are known even for activation functions of a single neuron, there are many open problems also in this field.

Coresets for Other Machine Learning Problems There are many problems in traditional machine learning with no coresets, including e.g. decision trees/forest with all their variants, as well as supported vector machines and other optimization functions such as RELU.

References

1. Agarwal, P.K., Har-Peled, S.: Maintaining the approximate extent measures of moving points. In: Proceedings of the 12th Soda, pp. 148–157 (2001)
2. Agarwal, P., Har-Peled, S., Varadarajan, K.: Approximating extent measures of points. J. Assoc. Comput. Mach. **51**(4), 606–635 (2004)
3. Agarwal, P., Har-Peled, S., Varadarajan, K.: Geometric approximation via coresets. Combinatorial Comput. Geom. **52**, 1–30 (2005)
4. Agarwal, P.K., Jones, M., Murali, T.M., Procopiuc, C.M.: A Monte Carlo algorithm for fast projective clustering. In: Proceeding ACM-SIGMOD International Conference on Management of Data, pp. 418–427 (2002)
5. Agarwal, P.K., Mustafa, N.H.: k-means projective clustering. In: Proceeding 23rd ACM SIGACT-SIGMOD-SIGART Symposium on Principles of Database Systems (PODS), pp. 155–165 (2004)
6. Agarwal, P.K., Procopiuc, C.M.: Approximation algorithms for projective clustering. In: Proceeding 11th Annual ACM-SIAM Symposium on Discrete Algorithms (SODA), pp. 538–547 (2000)
7. Agarwal, P.K., Procopiuc, C.M.: Approximation algorithms for projective clustering. J. Algorithms **46**(2), 115–139 (2003)
8. Agarwal, P.K., Procopiuc, C.M., Varadarajan, K.R.: Approximation algorithms for k-line center. In: European Symposium on Algorithms, pp. 54–63 (2002)
9. Aggarwal, A., Deshpande, A., Kannan, R.: Adaptive sampling for k-means clustering. In: Proceedings of the 25th approx, pp. 15–28 (2009)
10. Anthony, M., Bartlett, P.L.: Neural Network Learning: Theoretical Foundations. Cambridge University Press, Cambridge (1999)
11. Assadi, S., Khanna, S.: Randomized composable coresets for matching and vertex cover. CoRR. abs/1705.08242 (2017). Retrieved from http://arxiv.org/abs/1705.08242
12. Bachem, O., Lucic, M., Krause, A.: Coresets for nonparametric estimation-the case of dp-means. In: International Conference on Machine Learning (ICML) (2015)
13. Bădoiu, M., Clarkson, K.L.: Smaller core-sets for balls. In: Proceedings of the 14th soda, pp. 801–802 (2003)
14. Bădoiu, M., Clarkson, K.L.: Optimal core-sets for balls. Comput. Geom. **40**(1), 14–22 (2008)
15. Bambauer, J., Muralidhar, K., Sarathy, R.: Fool's gold: an illustrated critique of differential privacy. Vanderbilt J. Entertain. Technol. Law **16**, 701 (2013)
16. Barger, A., Feldman, D.: k-means for streaming and distributed big sparse data. In: Proceeding of the 2016 SIAM International Conference on Data Mining (sdm'16) (2016)
17. Batson, J., Spielman, D.A., Srivastava, N.: Twice-ramanujan sparsifiers. SIAM Rev. **56**(2), 315–334 (2014)
18. Baykal, C., Liebenwein, L., Gilitschenski, I., et al.: Data-dependent coresets for compressing neural networks with applications to generalization bounds. arXiv preprint arXiv:1804.05345 (2018)
19. Bentley, J.L., Saxe, J.B.: Decomposable searching problems I. static-to-dynamic transformation. J. Algorithms **1**(4), 301–358 (1980)
20. Blumer, A., Ehrenfeucht, A., Haussler, D., Warmuth, M.K.: Learnability and the vapnik-chervonenkis dimension. J. Assoc. Comput. Mach. **36**(4), 929–965 (1989)
21. Boutsidis, C., Zouzias, A., Mahoney, M.W., Drineas, P.: Randomized dimensionality reduction for k-means clustering. IEEE Trans. Inf. Theory **61**(2), 1045–1062 (2015)
22. Braverman, V., Feldman, D., Lang, H.: New frameworks for offline and streaming coreset constructions. arXiv preprint:1612.00889 (2016)
23. Charikar, M., Guha, S.: Improved combinatorial algorithms for facility location problems. SIAM J. Comput. **34**(4), 803–824 (2005)
24. Charikar, M., Guha, S., Tardos, É., Shmoys, D.B.: A constant-factor approximation algorithm for the k-median problem. J. Comput. Syst. Sci. **65**(1), 129–149 (2002)

25. Chazelle, B., Edelsbrunner, H., Grigni, M., Guibas, L., Sharir, M., Welzl, E.: Improved bounds on weak &egr;-nets for convex sets. In: Proceedings of the Twenty-Fifth annual ACM Symposium on Theory of Computing (STOC), pp. 495–504. ACM, New York (1993)
26. Chen, K.: On coresets for k-median and k-means clustering in metric and euclidean spaces and their applications. SIAM J. Comput. **39**(3), 923–947 (2009)
27. Choi, S., Kim, T., Yu, W.: Performance evaluation of ransac family. J. Comput. Vis. **24**(3), 271–300 (1997)
28. Clarkson, K.L.: Coresets, sparse greedy approximation, and the frank-wolfe algorithm. Assoc. Comput. Mach. Trans. Algorithms (TALG) **6**(4), 63 (2010)
29. Clarkson, K.L., Woodruff, D.P.: Numerical linear algebra in the streaming model. In: Proceedings of the 41st STOC, pp. 205–214 (2009)
30. Cohen, M.B., Elder, S., Musco, C., Musco, C., Persu, M.: Dimensionality reduction for k-means clustering and low rank approximation. In: Proceedings of the Forty-Seventh Annual ACM on Symposium on Theory of Computing, STOC 2015, pp. 163–172 (2015)
31. Cohen, M.B., Lee, Y.T., Musco, C., Musco, C., Peng, R., Sidford, A.: Uniform sampling for matrix approximation. In: Proceedings of the 2015 Conference on Innovations in Theoretical Computer Science, pp. 181–190. ACM, New York (2015) . http://doi.acm.org/10.1145/2688073.2688113
32. Dasgupta, A., Drineas, P., Harb, B., Kumar, R., Mahoney, M.W.: Sampling algorithms and coresets for ℓ_p-regression. In: Proceedings 19th Annual ACM-SIAM Symposium on Discrete Algorithms (SODA), pp. 932–941 (2008) . http://doi.acm.org/10.1145/1347082.1347184
33. Dasgupta, S., Schulman, L.J.: A two-round variant of em for gaussian mixtures. In: Proceedings of the Sixteenth Conference on Uncertainty in Artificial Intelligence, pp. 152–159 (2000)
34. Deshpande, A., Rademacher, L., Vempala, S., Wang, G.: Matrix approximation and projective clustering via volume sampling. In: Proceedings 17th Annual ACM-SIAM Symposium on Discrete Algorithms (SODA), pp. 1117–1126 (2006)
35. Drineas, P., Mahoney, M.W., Muthukrishnan, S.: Sampling algorithms for l_2 regression and applications. In: Proceeding of SODA 06 Proceedings of the Seventeenth Annual ACM-SIAM Symposium on Discrete Algorithm, pp. 1127–1136 (2006)
36. Edwards, M., Varadarajan, K.: No coreset, no cry: Ii. In: International Conference on Foundations of Software Technology and Theoretical Computer Science, pp. 107–115 (2005)
37. Effros, M., Schulman, L.J.: Deterministic clustering with data nets. In: Electronic Colloquium on Computational Complexity (ECCC), Report no. 050 (2004)
38. Epstein, D., Feldman, D.: Quadcopter tracks quadcopter via real-time shape fitting. IEEE Robot. Autom. Lett. **3**(1), 544–550 (2018)
39. Feigin, M., Feldman, D., Sochen, N.: From high definition image to low space optimization. In: Proceeding 3rd International Conference on Scale Space and Variational Methods in Computer Vision (SSVM 2011) (2011)
40. Feldman, D., Langberg, M.: A unified framework for approximating and clustering data. In: Proceeding 34th Annual ACM Symposium on Theory of Computing (STOC) (2011). See http://arxiv.org/abs/1106.1379 for fuller version
41. Feldman, D., Schulman, L.J.: Data reduction for weighted and outlier-resistant clustering. In: Proceeding of the 23rd Annual ACM-SIAM Symposium on Discrete Algorithms (SODA), pp. 1343–1354 (2012)
42. Feldman, D., Tassa, T.: More constraints, smaller coresets: constrained matrix approximation of sparse big data. In: Proceedings of the 21th ACM SIGKDD International Conference on Knowledge Discovery and Data Mining (kdd'15), pp. 249–258 (2015)
43. Feldman, D., Fiat, A., Sharir, M.: Coresets for weighted facilities and their applications. In: FOCS, pp. 315–324 (2006)
44. Feldman, D., Fiat, A., Segev, D., Sharir, M.: Bi-criteria linear-time approximations for generalized k-mean/median/center. In: Proceeding of 23rd ACM Symposium on Computational Geometry (SOCG), pp. 19–26 (2007)

45. Feldman, D., Monemizadeh, M., Sohler, C.: A ptas for k-means clustering based on weak coresets. In: Proceedings of the 23rd ACM Symposium on Computational Geometry (SoCG), pp. 11–18 (2007)
46. Feldman, D., Fiat, A., Kaplan, H., Nissim, K.: Private coresets. In: Proceedings of the Forty-First Annual ACM Symposium on Theory of Computing, pp. 361–370 (2009)
47. Feldman, D., Monemizadeh, M., Sohler, C., Woodruff, D.P.: Coresets and sketches for high dimensional subspace approximation problems. In: Proceedings of the Twenty-First Annual ACM-SIAM Symposium on Discrete Algorithms, pp. 630–649 (2010)
48. Feldman, D., Krause, A., Faulkner, M.: Scalable training of mixture models via coresets. In: Proceeding 25th Conference on Neural Information Processing Systems (NIPS) (2011)
49. Feldman, D., Sugaya, A., Rus, D.: An effective coreset compression algorithm for large scale sensor networks. In: 2012 ACM/IEEE 11th International Conference on Information Processing in Sensor Networks (IPSN), pp. 257–268 (2012)
50. Feldman, D., Sung, C., Rus, D.: The single pixel gps: learning big data signals from tiny coresets. In: Proceedings of the 20th International Conference on Advances in Geographic Information Systems, pp. 23–32 (2012)
51. Feldman, D., Sugaya, A., Sung, C., Rus, D.: Idiary: from gps signals to a text-searchable diary. In: Proceedings of the 11th ACM Conference on Embedded Networked Sensor Systems, p. 6 (2013)
52. Feldman, D., Schmidt, M., Sohler, C.: Turning big data into tiny data: constant-size coresets for k-means, PCA and projective clustering. In: Proceedings of the Twenty-Fourth Annual ACM-SIAM Symposium on Discrete Algorithms, pp. 1434–1453 (2013a)
53. Feldman, D., Schmidt, M., Sohler, C.: Turning big data into tiny data: constant-size coresets for k-means, pca and projective clustering. In: Proceedings of the Twenty-Fourth Annual ACM-SIAM Symposium on Discrete Algorithms, pp. 1434–1453 (2013b)
54. Feldman, D., Volkov, M., Rus, D.: Dimensionality reduction of massive sparse datasets using coresets. In: Advances in Neural Information Processing Systems (NIPS) (2016)
55. Feldman, D., Ozer, S., Rus, D.: Coresets for vector summarization with applications to network graphs. In: Proceedings of the 34th International Conference on Machine Learning, ICML 2017. Sydney, NSW, Australia, 6–11 August 2017, pp. 1117–1125 (2017). http://proceedings.mlr.press/v70/feldman17a.html
56. Feldman, D., Xiang, C., Zhu, R., Rus, D.: Coresets for differentially private k-means clustering and applications to privacy in mobile sensor networks. In: 2017 16th ACM/IEEE International Conference on Information Processing in Sensor Networks (IPSN), pp. 3–16 (2017)
57. Foster, I.: Designing and Building Parallel Programs. Addison Wesley Publishing Company, Reading (1995)
58. Funke, S., Laue, S.: Bounded-hop energy-efficient broadcast in low-dimensional metrics via coresets. In: Annual Symposium on Theoretical Aspects of Computer Science, pp. 272–283 (2007)
59. Har-Peled, S.: No coreset, no cry. In: Proceedings of the 24th IARCS Annual Conference on Foundations of Software Technology and Theoretical Computer Science (FSTTCS), pp. 324–335 (2004)
60. Har-Peled, S.: Coresets for discrete integration and clustering. In: 26th FSTTCS, pp. 33–44 (2006)
61. Har-Peled, S., Kushal, A.: Smaller coresets for k-median and k-means clustering. In: Proceedings of the 25th SODA, pp. 126–134 (2005)
62. Har-Peled, S., Mazumdar, S.: Coresets for k-means and k-median clustering and their applications. In: Proceedings of the 36th ACM Symposium on the Theory of Computing (STOC), pp. 291–300 (2004a)
63. Har-Peled, S., Mazumdar, S.: On coresets for k-means and k-median clustering. In: Proceeding of the 36th Annual ACM Symposium on Theory of Computing (STOC), pp. 291–300 (2004b)
64. Har-Peled, S., Varadarajan, K.R.: Projective clustering in high dimensions using coresets. In: Proceeding 18th ACM Symposium on Computational Geometry (SOCG), pp. 312–318 (2002)

65. Haussler, D.: Decision theoretic generalizations of the pac learning model. In: Proceedings of the 1st International Workshop on Algorithmic Learning Theory (ALT), pp. 21–41 (1990)
66. Haussler, D., Welzl, E.: Epsilon-nets and simplex range queries. In: Annual ACM Symposium on Computational Geometry (SOCG) (1986)
67. Hellerstein, J.: Parallel programming in the age of big data. In: GIGAOM blog. Nov. 9, 2008 (2008)
68. Hoeffding, W.: Probability inequalities for sums of bounded random variables. J. Am. Stat. Assoc. **58**(301), 13–30 (1963)
69. IBM: What is Big Data? Bringing Big Data to the Enterprise (2012). www.ibm.com/software/data/bigdata/. Accessed 3rd Oct 2012
70. Inaba, M., Katoh, N., Imai, H.: Applications of weighted voronoi diagrams and randomization to variance-based k-clustering. In: Symposium on Computational Geometry, pp. 332–339 (1994)
71. Indyk, P., Mahabadi, S., Mahdian, M., Mirrokni, V.S.: Composable core-sets for diversity and coverage maximization. In: Proceedings of the 33rd ACM Sigmod-Sigact-Sigart Symposium on Principles of Database Systems, pp. 100–108 (2014)
72. Joshi, S., Kommaraji, R.V., Phillips, J.M., Venkatasubramanian, S.: Comparing distributions and shapes using the kernel distance. In: Proceedings of the Twenty-Seventh Annual Symposium on Computational Geometry, pp. 47–56 (2011)
73. Langberg, M., Schulman, L.J.: Universal epsilon-approximators for integrals. In: Proceedings of the 21st ACM-SIAM Symposium on Discrete Algorithms (SODA), pp. 598–607 (2010)
74. Li, Y., Long, P.M., Srinivasan, A.: Improved bounds on the sample complexity of learning. J. Comput. Syst. Sci. **62**, 516–527 (2001)
75. Löffler, M., Phillips, J.M.: Shape fitting on point sets with probability distributions. In: European Symposium on Algorithms, pp. 313–324 (2009)
76. Mahajan, M., Nimbhorkar, P., Varadarajan, K.: The planar k-means problem is np-hard. In: International Workshop on Algorithms and Computation, pp. 274–285 (2009)
77. Mahoney, M.W.: Randomized algorithms for matrices and data. Found. Trends Mach. Learn. **3**(2), 123–224 (2011)
78. Matousek, J.: Approximations and optimal geometric divide-an-conquer. J. Comput. Syst. Sci. **50**(2), 203–208 (1995)
79. Matouaek, J.: New constructions of weak epsilon-nets. In: Proceedings of the Nineteenth Annual Symposium on Computational Geometry, pp. 129–135 (2003)
80. McLachlan, G., Krishnan, T.: The EM Algorithm and Extensions, vol. 382. Wiley, New York (2007)
81. Munteanu, A., Schwiegelshohn, C.: Coresets-methods and history: a theoreticians design pattern for approximation and streaming algorithms. KI-Künstl. Intell. **32**(1), 37–53 (2018)
82. Muthukrishnan, S., et al.: Data streams: algorithms and applications. Found. Trends Theor. Comput. Sci. **1**(2), 117–236 (2005)
83. Ostrovsky, R., Rabani, Y., Schulman, L.J., Swamy, C.: The effectiveness of lloyd-type methods for the k-means problem. In: 47th Annual IEEE Symposium on Foundations of Computer Science, 2006. FOCS'06, pp. 165–176 (2006)
84. Paul, R., Feldman, D., Rus, D., Newman, P.: Visual precis generation using coresets. In: 2014 IEEE International Conference on Robotics and Automation (ICRA), pp. 1304–1311 (2014)
85. Phillips, J.M.: Coresets and Sketches, Near-Final Version of Chapter 49 in Handbook on Discrete and Computational Geometry, 3rd edn. CoRR (2016). abs/1601.00617. Retrieved from http://arxiv.org/abs/1601.00617
86. Rosman, G., Volkov, M., Feldman, D., Fisher III, J.W., Rus, D.: Coresets for k-segmentation of streaming data. In: Advances in Neural Information Processing Systems (NIPS), pp. 559–567 (2014)
87. Segaran, T., Hammerbacher, J.: Beautiful Data: The Stories Behind Elegant Data Solutions. O'Reilly Media, Inc., Beijing (2009)
88. Sener, O., Savarese, S.: Active learning for convolutional neural networks: a core-set approach. Statistics **1050**, 27 (2017)

89. Shyamalkumar, N., Varadarajan, K.: Efficient subspace approximation algorithms. In: Proceedings of the 18th ACM-SIAM Symposium on Discrete Algorithms (SODA), pp. 532–540 (2007)
90. Tolochinsky, E., Feldman, D.: Coresets for Monotonic Functions With Applications to Deep Learning (2018). arXiv preprint:1802.07382
91. Tremblay, N., Barthelmé, S., Amblard, P.-O.: Determinantal Point Processes for Coresets (2018). arXiv preprint:1803.08700
92. Varadarajan, K., Xiao, X.: A near-linear algorithm for projective clustering integer points. In: Proceedings of the ACM-SIAM Symposium on Discrete Algorithms (SODA) (2012a)
93. Varadarajan, K., Xiao, X.: On the sensitivity of shape fitting problems. In: Proceedings of the 32nd Annual Conference on IARCS Annual Conference on Foundations of Software Technology and Theoretical Computer Science (FSTTCS), pp. 486–497 (2012b)

Chapter 3
A Family of Unsupervised Sampling Algorithms

Serge Guillaume and Frédéric Ros

3.1 Introduction

Defining a sample that behaves like the whole data set is a quite long-standing issue in data management. It has received fresh interest with the challenge of big data, characterized by an increase in the volume, velocity, and variety of the data. In this chapter, sampling aims to cope with the volume dimension.

The first attempt was the Lloyd algorithm designed in 1957, but only published in 1982 [40]. The goal was to find evenly spaced sets of points in subsets of Euclidean spaces, and partitions of these subsets into well-shaped and uniformly sized convex cells. It is closely related to the *k-means* algorithm, first proposed by James MacQueen in 1967 [44] and made popular by Hartigan [21], as both minimize the same objective function, called quantization distortion in signal processing. The main difference is that the Lloyd algorithm uses a Voronoi tessellation.

The Lloyd approach was generalized to any distribution, even with discrete components, by Linde, Buzo and Gray in 1980 [38]. Their technique does not involve any differentiation. This vectorial quantization yields an optimal codebook. The *LBG* algorithm is widely used in signal compression, either image or speech.

Recently, the concept of coreset, more precisely ε-coreset, was proposed [1]. The idea is to quantify the distortion of a given monotonic measure when computed on a sample instead of on the whole set. Extensive research has been carried out to generate such coresets in different frameworks [20].

S. Guillaume (✉)
UMR ITAP, Irstea, Montpellier, France
e-mail: serge.guillaume@irstea.fr

F. Ros
PRISME Laboratory, Orléans University, Orléans, France
e-mail: frederic.ros@univ-orleans.fr

F. Ros, S. Guillaume (eds.), *Sampling Techniques for Supervised or Unsupervised Tasks*, Unsupervised and Semi-Supervised Learning,
https://doi.org/10.1007/978-3-030-29349-9_3

Sampling and coresets have been applied to clustering. Clustering is an unsupervised task to organize, summarize, and finally understand the data. In the field of big data, clustering algorithms are becoming more and more sophisticated in order to deal with complex data of various shapes and densities [28]. Two challenging and connected problems arise with complexity: tuning and timing. Uniform sampling is the simplest and quickest way to proceed. Unfortunately, it requires very large sample sets to deal with shape and density variability. Smarter and more powerful algorithms have been proposed. These sampling algorithms are density or distance based and some of them combine the two notions under specific strategies. Density-based methods can be grouped in two main families for density estimation: space partition [27, 50] (e.g., grids, trees) and local estimation, using neighborhood or kernel functions [35]. Both are highly sensitive to parameters, cell definition for grids, bandwidth or neighborhood for local estimators. With an inappropriate setting, these methods are either likely to sample noise or to miss regions of interest. Distance-based clustering algorithms are used for sampling with a number of samples much greater than the number of clusters. The most famous representative of this family is the *k-means* [22]. Its sensitivity to initialization has been exhaustively investigated [4, 7, 68]. Single scan approaches have also been proposed such as *leader* [39] clustering. The results are highly dependent on the distance threshold, even with improved versions [59, 62].

Moreover, improvements in accuracy often conflict with time performance, and response time is of major concern nowadays for data processing algorithms. The increased computational cost limits the application of some of the above-mentioned algorithms to small or average size data sets. Several techniques have been investigated to address these challenges [41, 42, 59, 69]. The stratification concept has been proposed to speed up algorithms with quadratic or exponential time complexity. To overcome sensitivity to splitting, extensions have been proposed to work with non disjoint partitions, using replication techniques [43].

The objective of this work is to introduce a family of three sampling algorithms that are easy to tune, scalable, and yield a small size sample. The three of them combine density and distance and are based on the farthest-first traversal (*fft*) concept. They are iterative algorithms that add a new sample at each iteration. The new representative is chosen, in a given group which depends on the algorithm, as the farthest from the representative of the group. This ensures space coverage and allows for time optimization that makes them faster than any competing approach. They also share an interesting property: they yield a coreset. In [18], an *Efficient Coreset Construction via Adaptive Sampling* was proposed, involving density and distance concepts while biasing the random sampling. The coreset framework introduced the idea of approximation quantification: a subset is called a coreset of a whole set if solving the optimization problem on the subset gives an ε-approximate solution on the whole input set.

The three algorithms are also easy to tune as they have only one meaningful and dimensionless user parameter. The first two ones give the priority either to distance, *DIDES* [53] which stands for *DIstance and DEnsity based Sampling*, or density,

DENDIS [52] for, *DENsity and DIstance based Sampling*. This priority impacts the group in which the new representative is chosen as well as the stopping criterion. Their common parameter is called *granularity*: the lower its value the higher the sample size.

The third algorithm is called *ProTraS* [54], which stands for *Probabilistic Traversing Sampling*. It differs from the others as it is explicitly designed to produce a (k, ε)-coreset: the approximation level, ε, is its unique parameter and is also used as the stopping criterion. *ProTraS* manages the concepts of distance and density in a new probabilistic way: the representative is chosen in the group with the highest probability of cost reduction. This probability is computed according to the within group distance and to the representativeness of the sample item.

The three algorithms in the family have common properties but also some differences. They all achieve a trade-off between space coverage and density representation. They have little sensitivity to the initialization and can be fully deterministic, they are robust to noise and yield a sample size which mainly depends on the data structure.

The remaining of the chapter is organized as follows. An overview of unsupervised sampling method is provided in Sect. 3.2. The concepts shared by the algorithms in the family, *fft*, time optimization and relationship to coresets are introduced in Sect. 3.3. Then the three algorithms, *DIDES*, *DENDIS*, and *ProTraS*, are individually described in Sect. 3.4. Their common properties and differences are illustrated using synthetic data and analyzed in Sect. 3.5.

Finally, the main conclusions are stated in Sect. 3.6.

3.2 An Overview of Unsupervised Sampling

The objective of the sampling is to select a subset of the original data that behaves like the whole.

The first method to appear was random sampling, subject of many studies. The results are interesting from a theoretical point of view [10, 11, 26], but they tend to overestimate the sample size in non worst-case situations. A lot of work has been done to improve this basic algorithm. In our case, sampling is a pre-processing step for clustering and clustering is assessed according to compactness (or cluster homogeneity) and group separability. This calls for two basic notions: density and distance. Clusters can be defined as dense input areas separated by low density transition zones.

Sampling algorithms are based upon these two notions, one driving the process while the other is more or less induced. Strategies have also been developed to improve and speed up the sampling process. Several approaches benefit from a kd-tree implementation [29].

3.2.1 Density Biased Sampling

The main idea of density methods [8, 9, 49] is to add a bias according to space density, giving a higher probability for patterns located in less dense regions to be selected so as to ensure the representation of small clusters. These methods can be grouped in two main families: space partition and kernel estimation.

The simplest space partition based method is a grid with non overlapping cells. Palmer and Faloutsos [50] introduced the problem of non uniform sampling for clusters corresponding to skew size distributions. The space is divided into equally sized bins. The bins with a small number of patterns are considered with a higher probability. This method is easy to implement and has a reasonable time complexity which allows its use with large data sets. However, the results are obviously sensitive to cell definition and bias level: noise can be selected when the bias is too high. Some work has been done to adapt the grid to the data [27]. Finding new boundaries is not a trivial task and leads to a substantial increase in the computational cost. Trees can be seen as an extension of grids, where the cells have not the same size but are specific to a node. Nanopoulos et al. presented one of the pioneering studies for clustering and sampling using R-tree [48]. Points belonging to the same node are considered to have an identical probability of being selected. An approximate local density is thus given by the ratio of the node cardinality to the corresponding volume (hyperrectangle). The sampling is done according to the biased local densities. The results are highly sensitive to the splitting strategy as well as to some key parameters such as the stopping criterion. Algorithms like the minimum spanning tree [64] have been proposed to improve both cluster relevance and tractability. Kd-trees proved efficient for outlier detection [25]. The drawbacks induced by the binary split are well known: the partitions lack smoothness as similar data may be found on both sides of a given boundary.

The local density in each data point of the population can also be estimated using non parametric kernel or neighboring approaches.

The multidimensional kernel density estimator is defined as:

$$\hat{f}(x) = \frac{1}{nh_1 \ldots h_d} \sum_{i=1}^{n} \left[\prod_{j=1}^{d} K\left(\frac{x_j - x_j^i}{h_j} \right) \right]$$

where d is the space dimension, n the set size, $K(\cdot)$ the kernel, and h the bandwidth.

Under weak conditions ($h_1, \ldots, h_d$ decrease when n increases) the estimate converges in probability to the true density. A lot of work has been dedicated to kernel shape and its impact on results [5, 45, 46], but the most influential parameter is the bandwidth. It controls the smoothness of a density estimate. If improperly defined, it can lead to rough estimations.

The general expression for neighboring density estimation is the following:

$$\hat{f}(x) = \frac{k}{nV}$$

where V is the volume surrounding x and k the number of items located in V. The estimate can be reached using two methods: either setting V and computing k or finding the k nearest neighbors of x and then deducing the corresponding volume.

Once the local densities have been estimated, they are used to bias the sampling process. In the work by Kollios et al. [35], the kernel is that of Epanechnikov [16] and the bias is as follows:

$$\frac{s}{\sum \hat{f}(x)} \hat{f}(x)^b$$

where s is the desired sample size and b the bias parameter.

If $b = 0$, the process reduces to a random sampling $\left(\frac{s}{n}\right)$. Otherwise, if $b > 0$ (respectively $b < 0$) high density regions are sampled at a higher (lower) rate.

Although extensively studied, these approaches are rather difficult to parameterize. With an inappropriate setting, these methods are either likely to sample noise or to miss some regions of interest. Moreover, clusters' shapes are not taken into account, nothing ensures that they are preserved in the sample set. This is an intrinsic limitation of these methods. Moreover, they usually require significant storage capacity and have a high computational cost.

3.2.2 Distance-Based Sampling

Density is an important cluster feature. Distance is the other notion involved in cluster definition, as it is used to measure similarity and proximity between patterns. For this reason, it is widely used in clustering and sampling algorithms.

The most popular algorithm representative of this family is the *k-means* [22], and its robust version called *k-medoids* [32]. This simple and powerful algorithm can be used for sampling large data sets. It remains one of the most influential and studied clustering algorithms [34, 41, 69]. Some work has been done about its computational efficiency [12], but most studies deal with the quality of the initial partition. They address the well-known *k-means* shortcoming: its sensitivity to initialization [4, 7, 68]. Different approaches based on sampling or condensing techniques, including evolutionary algorithms [23, 47], have been investigated.

The original version is limited as it only produces spherical clusters. It has been enriched to deal with more complex data and to yield overlapping clusters. The fuzzy extension [6] is widely used, while the possibilistic version [36] does not constrain the sum of the membership degrees to be 1.

k-means has been successfully used as a pre-processing sampling step for sophisticated and expensive techniques such as hierarchical approaches or support vector machine algorithms (SVM) [61, 65]. It is run with $k = s$, the number of representatives, such as: $c \ll s \ll n$, c being the unknown number of clusters.

While the *k-means* is an iterative algorithm, whose convergence is guaranteed, some single data-scan distance-based algorithms have also been proposed, such as *leader* clustering [39]. Each pattern, x is assigned to a leader, l, if $d(x,l) \leq t$, t being a predefined threshold. If there is no leader in x neighborhood, x becomes a new leader. The results are sensitive to the threshold, and to the initial pattern order. This basic version has been improved [58, 62]. In the latter various thresholds are used to yield clusters of different sizes. First the *k-means* is run on a small random sample of the original data, with $k \gg c$ groups. The centers (means) are $m_1, \ldots, m_k$. The threshold distance for a new leader, x, is computed from its two nearest means, m_i and m_j, as follows:

$$t(x) = \lambda\,(a - b), \quad 0 < \lambda \leq 1, \text{ where } a = \frac{\|m_i - m_j\|}{2} \text{ and } b = (x - m_i)\frac{m_i - m_j}{\|m_i - m_j\|}.$$

The *leader* method is a pure distance-based algorithm that does not account for density. The *mountain method* [13, 66] can be considered as an improved leader approach in which local density is also taken into account. It finds first the most representative pattern in order to be less dependent on the presentation order. Like the *k-means*, *leader* algorithms can be used for sampling with a threshold $t' \ll t$ to give a number of representatives $s \gg c$.

The pioneering versions of distance-based methods, such as the leader family approaches, are simple and fast but clearly limited. When improved, by taking density into account [55], they become more relevant but their overall performance depends on the way both concepts are associated and on the increased computational cost.

The *mountain method* proposed by Yager and its modified versions [67] are good representatives of hybrid methodologies as well as the recent work proposed by Feldman et al. [18]. Density is managed by removing from the original set items already represented in the sample.

3.2.3 Stratification Strategies

The stratification concept has been proposed to speed up algorithms with quadratic or exponential time complexity. The idea is to divide the set into subsets, called strata, that are processed, or sampled in our case, independently. The final sampling is built from the union of the samples of each data set.

The main drawback of stratification methods stems from the splitting: it is done randomly, with no prior knowledge about the data distribution. This does not ensure that the most informative patterns are selected in each subset. This highlights the importance of the final aggregation step. When each subset has been sampled, the sampling can be reiterated on the union of the selected patterns to keep only the most representative ones.

In the earliest versions the strata had the same size, collectively exhaustive and mutually exclusive. As an example, Bagged clustering [14, 37] consists in running a cluster method on each subset with the same number of clusters.

Extensions have been proposed to work with non disjoint partitions, using replication techniques [43]. These methods take density into consideration. Dense areas are obviously represented in the strata while regions corresponding to noisy data are likely to be diluted in the whole set of strata. This way, they have less opportunity to be represented in the final set.

Stratification can be combined with boosting strategies [19]. The former speeds up the process while the latter improves the sampling or clustering relevance.

Reservoir algorithms [63] are incremental and can be seen as a special case of stratification approaches. They have been proposed to deal with dynamic data sets, like the ones to be found in web processing applications.

Many variants exist. As examples [2] is adapted to handle heterogeneous data distribution and [15] considers sampling with and without replacement.

Some studies have been done to adapt the size of the reservoir [3] and to propose appropriate aggregation strategies. Density is explicitly managed in [33] using the weighted *k-means*.

This short survey shows that sampling for clustering techniques have been well investigated. Both concepts, density and distance, and the methods have reached a good level of maturity. Some work deal with algorithm computational efficiency [59], but only a few papers study the sample size [51, 60]. As far as we know, the question of parameter tuning has not really been addressed. The new challenge is to take the best of the available techniques and combine them in order to propose a self-adaptive, data independent, and tractable algorithm able to process various kinds of large data sets with a standard setting.

3.3 Common Concepts Shared by the Three Algorithms

The algorithms in the family are based on three concepts which are introduced in this section. The farthest-first traversal concept comes to select the new representative as the farthest from the representative in a given group. This distance property can be used to optimize the algorithm thanks to the triangular inequality. The third common point is that these algorithms yield a coreset.

3.3.1 The Farthest-First Traversal Item as a Representative

The algorithms introduced in this chapter are iterative algorithms that add a new item to the sample at each loop, until a stopping criterion is met. They share the way the new representative is chosen: this process, based on the distance, is also known as the farthest-first traversal (*fft*) algorithm. This concept has been used 30 years ago to initialize the *k-means* algorithm [24, 56]. Sensitive to outliers, it has inspired Arthur and Vassilvitskii [4] to propose *kmeans++*: new seeds are randomly chosen with a probability proportional to their distance to already chosen ones.

The proposed algorithms can be considered as *fft* improvements. The concept is summarized in Algorithm 1.

Algorithm 1 *fft* algorithms: representative selection

1: Input: $T = \{x_i\}, i = 1 \ldots, n$
2: Output: $S = \{y_j\}, T(y_j), j = 1, \ldots, s$
3: Select an initial pattern $x_{init} \in T$
4: $S = \{y_1 = x_{init}\}, s = 1$
5: **repeat**
6:     **for all** $x_l \in T \setminus S$ **do**
7:         Find $d_{near}(x_l) = \min_{y_k \in S} \; d(x_l, y_k)$
8:         $T(y_k) = T(y_k) \cup \{x_l\}$ {Set of patterns represented by y_k}
9:     **end for**
10:    **for all** $y_k \in S$ **do**
11:         Find $d_{max}(y_k) = \max_{x_m \in T(y_k)} \; d(x_m, y_k)$
12:         Store $d_{max}(y_k), x_{max}(y_k)$ {where $d_{max}(y_k) = d(x_{max}(y_k), y_k)$},
13:    **end for**
14:    {Select the farthest pattern from the representative in a given group.}
15:    {The group, w, depends on the proposed algorithm: $y_* = x_{max}(y_w)$.}
16:    $S = S \cup \{y_*\}$
17: **until** Stopping condition is met
18: **return** $S, T(y_j), j = 1, \ldots, s$

Let $T = \{x_i\}$ be the input set of n multidimensional data, and $S = \{y_j\}$ the size-s sample to be built, $S \subset T$. The set of patterns represented by y_k is: $T(y_k) = \{x_i \mid d(x_i, y_k) = \min_j d(x_i, y_j)\}$.

The first pattern can be randomly chosen or it can be computed as the farthest, depending on the selected distance, from the minimum value in each input space dimension. The latter makes the algorithm fully deterministic. After the initialization phase, the set S only counts this initial pattern, x_{init} (lines 3–4).

The main loop (lines 5–17) includes two steps. First, each unselected pattern, $x \in T \setminus S$, is attached to the closest selected one in S (lines 6–9). At the first step, $T(y_1) = T \setminus \{x_{init}\}$. Then, for each set $T(y_k)$, the algorithm searches for the farthest attached pattern located at the distance $d_{max}(y_k)$ (lines 10–13).

The next selected representative, x_w, is the farthest item chosen in a given group. So, boundary patterns are first chosen instead of inner ones. This way, the selected set spans the whole input space.

3.3.2 Optimization

Distance-based algorithms have an usual complexity of $O(n^2)$. This is not the case for the proposal as the *fft* concept allows for some optimization. Many distance computations can be avoided thanks to the algorithm structure itself and by using the triangular inequality.

For each of the s iterations, the first loop, lines 6–9, computes $(n - s)n$ distances while the second one, lines 10–13, calculates n more ones.

3.3.2.1 Reducing Time Complexity

These two loops can be combined in a single one, lines 2–12 in Algorithm 2. This allows for only computing $n - s$ distances to the new representative, y_*, at each of the s iterations.

Algorithm 2 The first two loops are combined into a single one

1: **while** (ADD==TRUE) **do**
2:     **for all** $x_l \in T \setminus S$ **do**
3:         Compute $d = d(x_l, y_*)$
4:         **if** $(d < d_{near}(x_l))$ **then**
5:             $T_{y_*} = T_{y_*} \cup \{x_l\}$, $T_{y(x_l)} = T_{y(x_l)} \setminus \{x_l\}$
6:             $d_{near}(x_l) = d$, $y(x_l) = y_*$
7:         **end if**
8:         **if** $(d > d_{max}(y_*))$ **then**
9:             $x_P = x(x_s)$, $Y_P = y_*$
10:             $d_{max}(y_*) = d$, $x(y_*) = x_l$
11:         **end if**
12:     **end for**
13:     Find a new representative y_*.
14: **end while**

The complexity is then $O(ns)$, with $s \ll n$.

The number of distances to be computed is:

$$T = \sum_{l=s}^{n} (l - 1) = \frac{n\,(n-1)}{2} - \frac{s\,(s-1)}{2} \tag{3.1}$$

The spatial complexity for this time optimization can be considered as reasonable: $n + 2s$ distances between the representatives are stored: n $d_{near}(x)$and s $d_{max}(y)$ as well as the corresponding elements, y for $d_{near}(x)$, and x for $d_{max}(y)$.

3.3.2.2 Using the Triangle Inequality

A given iteration only impacts a part of the input space, meaning the neighborhood of the new representative. Moreover as the process goes on, the corresponding induced volume decreases. This may save many distance calculations.

Algorithm 3 The optimized version of the *fft* algorithm

1: Input: $T = \{x_i\}, i = 1 \ldots, n, g_r$
2: Output: $S = \{y_j\}, \{T_{y_j}\}, j = 1, \ldots, s$
3: ADD=TRUE, $W_t = n\ g_r$
4: Select an initial pattern $x_{init} \in T$
5: $S = \{y_1 = y_* = x_{init}\}, s = 1$
6: $d_{near}(x_i) = \infty, i = 1 \ldots, n$
7: **while** (ADD==TRUE) **do**
8: $F = \{T_{y_j} | d(y_j, y_*) \geq 2\ d_{max}(y_j)\}$
9: **for all** $x_l \in T \setminus \{S \cup F\}$ **do**
10: **if** ($d_{near}(x_l) > 0.5\ d(y(x_l), y_*)$) **then**
11: Compute $d = d(x_l, y_*)$
12: **if** ($d < d_{near}(x_l)$) **then**
13: $T_{y_*} = T_{y_*} \cup \{x_l\}, T_{y(x_l)} = T_{y(x_l)} \setminus \{x_l\}$
14: $d_{near}(x_l) = d, y(x_l) = y_*$
15: **end if**
16: **if** ($d > d_{max}(y_*)$) **then**
17: $x_P = x(x_s), Y_P = y_*$
18: $d_{max}(y_*) = d, x(y_*) = x_l$
19: **end if**
20: **end if**
21: **end for**
22: Find a new representative y_*
23: **end while**
24: **return** $S, T_{y_k} \forall k \in S$

When a new representative in S has been selected, y_*, the question is: should a given initial pattern, x_i, be attached to y_* instead of remaining in T_{y_j}? The triangular inequality states: $d(y_j, y_*) \leq d(x_i, y_j) + d(x_i, y_*)$. And, $x_i \in T_{y_*} \iff d(x_i, y_*) < d(x_i, y_j)$. So, if $d(y_j, y_*) \geq 2\ d(x_i, y_j)$, x_i remains in T_{y_j}, no change needs to be made. Only two distances are needed to check the inequality, and discard any further calculations in the case of no change. In our algorithm, there is no need to check this inequality for all the initial patterns. For each representative, y_k, $d_{max}(y_k)$ is stored. If $d(y_k, y_*) \geq 2\ d_{max}(y_k)$, meaning the furthest initial pattern from y_k remains attached to y_k, this also holds $\forall x_i \in T_{y_k}$. Then, these representatives and their attached patterns are not concerned by the main loop of the algorithm (Algorithm 3, line 8–9). When this is not the case, the same triangle inequality provides a useful threshold. All $x_i \in T_{y_k}$ with $d_{near}(x_i) \leq 0.5\ d(y_j, y_*)$ remain attached to T_{y_k} (line 10).

To take advantage of the triangular inequality properties, the number of distances between representatives to be stored is $s(s-1)/2$.

The optimized version of the sampling algorithm is shown in Algorithm 3.

In the case illustrated in Fig. 3.1 $y_* = y_4$ is the farthest item in *Group 1*, x_1, and the closest representative from y_1 is y_3. The limits of the new group, *Group 4*, are plotted in dashed lines. *Group 3* is reduced and x_i becomes the farthest element from y_3, and it is now labeled as x_3. Similarly x_j becomes x_4.

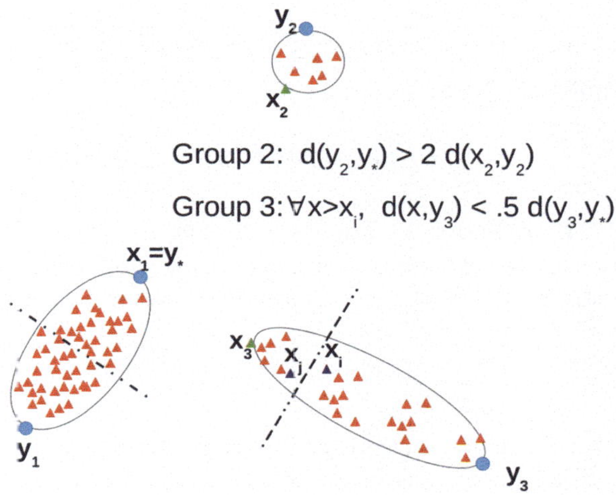

Fig. 3.1 Optimization illustration

3.3.2.3 Estimating the Number of Computed Distances

The number of computed distances cannot be rigorously defined as it depends on the data, but it can be however roughly estimated under some weak hypothesis. Each iteration of this distance based algorithm impacts only the neighborhood of the new representative. Let k be the number of neighbors to consider. The number of distances to be calculated is $(n-1)$ at the first step, then the number of representatives to take into account is $min(k,s)$, and the number of patterns for which the distance to the representatives has to be computed is only a proportion, δ, of the set of the attached ones as the others are managed by the triangular inequality properties. A value of $\delta = 0.5$ seems to be reasonable. This means that a high proportion of representatives are concerned at the starting of the algorithm but the process then becomes more and more powerful when s increases compared to k. The real number of computed distances can be estimated as follows:

$$C = (n-1) + \sum_{i=s}^{n-1} \sum_{l=1}^{min(k,s-i)} \delta * |T_{y_l}(i)| \tag{3.2}$$

where $|T_{y_l}(i)|$ is the number of patterns attached to representative l when i representatives are selected.

To approximate C, one can consider that on average the representatives have a similar weight $\forall y, |T_{y_l}(i)| \approx n/i$. When the two cases, $i \leq k$ and $i > k$, are developed, the approximation becomes:

$$C = (n-1) + \delta \left(\sum_{i=2}^{k+1} (i-1)\,\frac{n}{i} + \sum_{i=s}^{n-k-2} k\,\frac{n}{i} \right)$$

As $\sum_{i=2}^{k+1}(i-1)\,(n/i) \le \sum_{i=2}^{k+1}(i)\,(n/i)$ and $\sum_{i=s}^{n-k-2} k\,\frac{n}{i} \le \sum_{i=s}^{n-k-2} k\,\frac{n}{s}$, an upper bound of C can be defined as follows:

$$C \le (n-1) + \delta\left(n(k-1) + k\frac{n}{s}(n-k-2-s)\right) \tag{3.3}$$

As an illustration, using $n = 20{,}000$, $s = 250$, $k = 10$, and $\delta = 0.6$, the ratio of the number of computed distances to the number of distances that would have been calculated without the optimization procedure, as given in Eq. (3.1), is:

$$D = \frac{C}{T} \le 5\%$$

This estimation is clearly confirmed by the experiments.

Under some reasonable assumptions, it can be estimated that most of distance calculations can be saved by judiciously using the triangle inequality. This optimization makes these algorithms very tractable.

3.3.3 Relationship to Coresets

The challenge of big data has aroused a new interest in sampling techniques. But the idea is not new: vector quantization was introduced in the field of signal and image processing to summarize a data distribution by a finite number of vectors [38].

More recently, the coreset framework introduced the idea of approximation quantification: a subset is called a coreset of a whole set if solving the optimization problem on the subset gives an ε-approximate solution on the whole input set.

This section aims to investigate whether there is a relationship between the proposed *fft* algorithms and coresets.

This concept was initially analyzed by Agarwal et al. [1] for the geometric approximation of point sets. Given a monotone measure function, μ, i.e., for $S \subseteq T$, $\mu(S) \le \mu(T)$, and given $\varepsilon > 0$, $S \subseteq T$ is an ε-coreset for T with respect to μ, if $(1-\varepsilon)\mu(T) \le \mu(S)$. Typical measures include statistics about the set itself such as diameter, width or the geometric shape enclosing T, e.g., the smallest enclosing ball characteristics such as radius or volume.

They proved that this approximation can be obtained using a sample whose size is independent of the number of points and only dependent on ε.

This concept has been extended to clustering applications [20]. The authors proposed the following definition.

Definition 3.1 A set S, of s items, is an (k, ε)-coreset for a set T, of $n > s$ items if:

$$(1-\varepsilon)Cost_T(C) \le Cost_S(C) \le (1+\varepsilon)Cost_T(C) \tag{3.4}$$

where $C = \{c_1, \ldots, c_k\}$ is a set of k centers.

Let $c_i^* \in C$ be the closest center for a given $x_i \in T$: $d(x_i, c_i^*) = \min_{m \in 1,\ldots,k} d(x_i, c_m)$.

Similarly, let $c_j^{*'} \in C$ be the closest center for a given $y_j \in S$: $d(y_j, c_j^{*'}) = \min_{m \in 1,\ldots,k} d(y_j, c_m)$.

With the *k-means* algorithm, the two costs are:

- $Cost_T(C) = \sum_{i=1}^{n} d(x_i, c_i^*)$
- $Cost_S(C) = \sum_{j=1}^{s} w_j\ d(y_j, c_j^{*'})$, with $w_j = |T(y_j)|$, i.e., the number of items from T whose closest point in S is y_j.

When this definition only holds for the optimal number of centers, k, S is called a weak coreset for T, otherwise, if it holds for any set C, it is called a strong coreset for T.

Theorem 3.1 *The proposed fft algorithms yield a* (k, ε)*-coreset.*

Proof As y_j also belongs to T, let $d(y_j, c_i^*)$ be the distance between the representative and its closest center computed from the whole set T.

One obtains $\forall j \in S$: $d(y_j, c_i^*) > d(y_j, c_j^{*'})$ if $c_i^* \neq c_j^{*'}$ otherwise $d(y_j, c_i^*) = d(y_j, c_j^{*'})$ then $\sum_{j=1}^{s} w_j\ d(y_j, c_j^{*'}) \leq \sum_{j=1}^{s} w_j\ d(y_j, c_i^*)$

The triangle inequality yields: $d(y_j, c_i^*) \leq d(x, y_j) + d(x, c_i^*)$

For a given $j \in S$: $w_j\ d(y_j, c_i^*) \leq w_j\ d_j + \sum_{l=1}^{w_j} d(x_l, c_i^*)$, $d_j = d_{max}y(j)$ being the maximum inner distance for group j.

Considering the whole set of the representatives: $\sum_{j=1}^{s} w_j\ d(y_j, c_j^{*'}) \leq \sum_{j=1}^{s} w_j\ d(y_j, c_j^*) \leq \sum_{j=1}^{s} w_j\ d_j + \sum_{i=1}^{n} d(x_i, c_i^*)$ then

$$Cost_S(C) \leq \sum_{j=1}^{s} w_j\ d_j + Cost_T(C)$$

The quantity $\sum_{j=1}^{s} w_j\ d_j$ is the sampling cost, or the quantization distortion.

The triangle inequality also gives: $d(x, c_j^{*'}) \leq d(x, y_j) + d(y_j, c_j^{*'})$ then $d(y_j, c_j^{*'}) \geq d(x, c_j^{*'}) - d(x, y_j)$

For all the j groups, it gives:

$\sum_{j=1}^{s} w_j\, d(y_j, c_j^{*'}) \geq \sum_{i=1}^{n} d(x_i, c_i^*) - \sum_{j=1}^{s} w_j\, d_j$

Meaning:

$$Cost_S(C) \geq Cost_T(C) - \sum_{j=1}^{s} w_j\, d_j$$

The final relation between the two costs is:

$$Cost_T(C) - \sum_{j=1}^{s} w_j\, d_j \leq Cost_S(C) \leq Cost_T(C) + \sum_{j=1}^{s} w_j\, d_j \tag{3.5}$$

Dividing Eq. (3.5) by $Cost_T(C)$, and then multiplying by $Cost_T(C)$, yields:

$$\left(1 - \frac{\sum_{j=1}^{s} w_j\, d_j}{Cost_T(C)}\right) Cost_T(C) \leq Cost_S(C) \leq \left(1 + \frac{\sum_{j=1}^{s} w_j\, d_j}{Cost_T(C)}\right) Cost_T(C)$$

which is Eq. (3.4) with $\varepsilon = \frac{\sum_{j=1}^{s} w_j\, d_j}{Cost_T(C)}$. □

S is thus a (k, ε)-coreset for T, with ε equals the ratio of the sampling cost to the whole cost. As there is no assumption about C, S is a strong coreset for T.

The sampling cost and, consequently, ε has a monotonic evolution with respect to the granularity. A lower granularity tends to yield a higher sample size and adding a new item to S decreases both w and d. This monotonic behavior may be locally unchecked due to the random initialization and to the inner control mechanisms of the two algorithms.

3.4 The Three Algorithms in the Family

Besides tractability tuning is of prime concern to the end user. The three proposed algorithms are driven by a unique, and meaningful parameter. The sample size is not a priori defined, it depends on the data structure.

The first two ones are called *DIDES* [53] and *DENDIS* [52] and combine distance and density concepts. Their unique parameter is called granularity and labeled g_r. It impacts the S size, in that the lower the $granularity$ the higher the number of representatives. It is data independent, and is combined with the whole set cardinality, n, to define a threshold, $n\, g_r$. This threshold has a different meaning for the two algorithms: it is the minimum size of a cluster one wants to have a representative in *DIDES* while in *DENDIS* it is the minimum size for a group to be split.

The last one, *ProTras* [54], is only driven by the sampling cost.

3.4.1 DIDES: Distance First

In *DIDES*, distance is the dominant criterion and the new representative is the farthest item from all the already selected ones. This ensures space coverage. The threshold th is the minimum size, in the initial set, T, for a cluster one wants to have representatives in S. A representative with fewer than th patterns attached is called a poor representative. When the proportion of T whose representative is a poor representative is high enough, the input space is homogeneously covered. Then, the d_{max} evolution curve can be modeled to define the stopping criterion as a distance threshold. Density is managed in a post-processing step to discard outliers and consider the representation of connected areas.

3.4.1.1 Beginning of the Algorithm

The selection of a new representative, line 14 of Algorithm 1, is as follows:

$$y_w = \arg\max_{y_k \in S} d_{max}(y_k), \quad y_* = x_{max}(y_w), \quad MaxDmax = d_{max}(y_w)$$

The new representative is chosen as the farthest from the representatives in the group with the maximum inner distance. Boundary patterns are first chosen instead of inner ones.

Figure 3.2 illustrates the first steps of our algorithm (blue triangles) with well-structured data including clusters of various shapes and densities, and then shows the input space is correctly spanned (red symbols).

The stopping criterion is not defined yet. Rather than the common number of samples we introduce step by step, in the following sections, an adaptive threshold on the $MaxDmax$ distance.

3.4.1.2 The $MaxDmax$ Evolution Properties

The $MaxDmax$ criterion used for selecting a new representative is monotonically decreasing. Unfortunately, the evolution curve, shown in blue in Fig. 3.3, does not indicate any break-point that would serve as a stopping criterion. A proportion in the decrease, with respect to the initial value, is not stable enough: the optimal threshold is really data dependent.

Fig. 3.2 *DIDES*: first steps of the algorithm. Data (yellow)—First selected items (blue triangles)—Sample points (red)

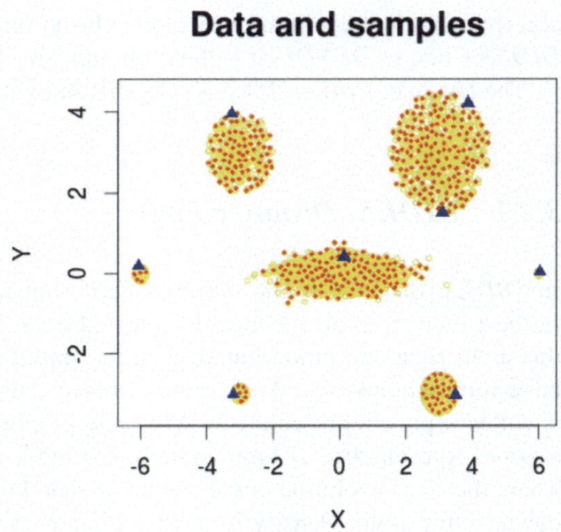

Fig. 3.3 Distance evolution curves for the data in Fig. 3.2: $MaxDmax$ (blue), $MinDmax$ (red), $MinDmaxW$ (dashed green) is similar to $MinDmax$ when the small cluster located at (6, 0) has been removed

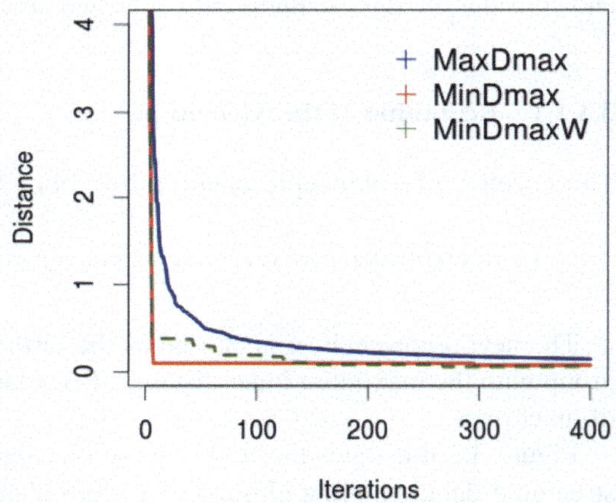

In the flat area, the number of samples is highly sensitive to the value of the $MaxDmax$ criterion. To ensure the representativeness, a cautious attitude would be to set a small threshold value, but this would lead to a number of selected samples higher than required to obtain the expected behavior. The challenge comes to estimate an acceptable trade-off.

Based on the common definition for a cluster, i.e., a dense area separated from another dense area by a sparsely populated zone, one needs to make sure that all the clusters are represented by the selected data. The narrowest cluster is correctly identified when the minimum of the d_{max} for all the representatives, hereafter called $MinDmax$ becomes equal to the smallest cluster dimension which is unknown. Let's make clear the item corresponding to the $MinDmax$ is never selected as a representative, only the one corresponding to the $MaxDmax$ is added to the sample set.

Once this value has been reached, the algorithm samples only the inner cluster structures. Defining the distance threshold according to the d_{min} evolution would ensure a perfect data coverage. However, in the case of singular clusters of small sizes, it would yield an over representation since the same distance needed for small clusters would also be used for the large ones. This is illustrated by the $MinDmax$ evolution shown in red in Fig. 3.3.

When the small cluster located at (6, 0) in Fig. 3.2 has been removed, this curve, $MinDmaxW$, is likely to provide valuable information, as shown with the dashed green line in Fig. 3.3.

As no threshold can be identified, its estimation for a given data set is based on a $MaxDmax$ evolution model. But a new issue arises: how to characterize the evolution stage that allows for building an accurate model?

Let $MinCard$ be the minimum size, in the initial set T, for a cluster one wants to have representatives in S. y_k is labeled as a poor representative if $|T(y_k)| < MinCard$. This parameter can be set according to the initial data set size and the desired granularity, $MinCard = n\ g_r$.

Let P be the proportion of initial data represented by poor representatives:

$$P = \frac{1}{n} \sum_{k=1}^{s} \left\{ |T(y_k)|,\ |T(y_k)| < MinCard \right\}$$

P evolves from 0 to 1, when s is close to n. P is a function of s, it is nearly monotonically increasing. It is getting positive as soon as an outlier, part of an isolated small cluster or noise, has been selected as a representative. The idea is then to threshold P to model the $MaxDmax$ evolution curve. It is expected that a wide range of P value yield a similar threshold.

When P has reached its desired value, meaning that this proportion of T is represented in S by small sets, all the small clusters in T are represented in S according to the $MinCard$ parameter.

3.4.1.3 Estimating a Lower Bound of the Threshold

The $MaxDmax$ curve has reached the flat area, and its evolution can be modeled by a power function $f(x) = ax^p$. The representatives are now numerous enough to make a robust estimation of the two parameters thanks to a least squares minimization. The so-called asymptote value is used as a lower bound of the threshold. The asymptote is characterized by small variations in $f(x)$; it is reached for s_a such that:

$$\frac{f(s_a)}{f(s_a + 1)} \geq (1 - \epsilon) \tag{3.6}$$

where ϵ is a small positive value set to 0.001.

Taking the logarithm of Eq. (3.6) yields:

$$\ln\left(\frac{x}{x+1}\right) \geq \frac{\ln(1-\epsilon)}{p} \text{whichgives} : s_a = -\frac{1}{1-\mathrm{e}^{\frac{\ln(1-\epsilon)}{p}}}$$

Finally, $th_a = f(s_a) = as_a^p$.

3.4.1.4 An Adaptive Stopping Criterion

This lower bound of the distance threshold th_a has been computed with a generic configuration, defined by three hidden parameters: $P = 0.2$, $\epsilon = 0.001$, and $g_r = 0.01$. With this configuration, shapes in the original space are well represented in the selected sample, whatever the data distribution and the noise amount.

To define the distance threshold that serves as a stopping criterion for the algorithm, the data as well as the granularity input parameter are now considered.

The selected sample is analyzed according to a twofold point of view: the $d_{max}(y_k)$ and the cardinality of the set attached to the sample, $|T(y_k)|$ for representative, y_k. Both distributions are used to define the threshold.

As the space is correctly covered, this is guaranteed by the P proportion for $g_r = 0.01$, outliers or noise representatives selected according to the distance criterion are now quite isolated, meaning both $d_{max}(y_k)$ and $|T(y_k)|$ are low. Let $S_t = \{y_k \mid |T(y_k)| < MinCard \ \& \ d_{max}(y_k) > th_a\}$ be a subset of representatives, th_a is the lower bound of the threshold and $MinCard$ is now computed with the user granularity. S_t includes small cluster representatives as well as outliers. The objective is to define a threshold that allows for keeping the former while removing the latter. The desired threshold, th, can be merely set as the average d_{max} over S_t:

$$th = \underset{y_k \in S_t}{mean} \ d_{max}(y_k) \tag{3.7}$$

Then, the algorithm iterates until this threshold is reached. If the user granularity is higher than 0.01, no more iteration may be needed.

3.4.1.5 Post-Processing: Density Area Management

As the selection is only done according to the distance from already selected items, density has to be taken into account with a twofold objective: remove noise points that might have been selected, especially during the first steps of the process and ensure dense area representativeness in the sample.

The first step is to remove noisy representatives. A threshold can be simply defined from the S_t set introduced in the previous section:

$$T_n = \underset{y_k \in S_t}{mean} \ |T(y_k)| \tag{3.8}$$

All the representatives with $|T(y_k)| < T_n$ are removed from S. The others are kept, meaning that even if they do not have $MinCard$ attached items they are not considered as noise.

Then density specific management is based on groups defined as connected areas. How to define a connected group of representatives? Representatives that belong to the same group are closer, one to each other, than representatives that belong to different groups. So, the group can be defined as the set of reachable points, from one another, within a given distance. The problem is now to set the threshold. It is based upon the distribution of distances between the representatives and their nearest neighbor, $d_{1nn}(y_k)$, and also on the $d_{max}(y_k)$ distribution. It is worth mentioning the $d_{1nn}(y_k)$ distances are updated at each iteration without any extra calculation.

The $d_{1nn}(y_k)$ distribution is filtered to only consider the potential connected pairs. As the regions of space containing data are homogeneously covered, neighbors separated by a distance higher than twice the $MaxDmax$ cannot be part of the same group. So, these values, $d_{1nn}(y_k) > 2\ MaxDmax$, are not taken into account to compute the basics statistics of the distribution: the average, $\overline{d}$, and standard deviation, σ, of the d_{1nn} distribution, and let $MaxD1nn$ be the maximum of the distribution. The reachable threshold is defined as:

$$d_r = \min\left(\overline{d} + 2\sigma, MaxD1nn\right) \tag{3.9}$$

The density is considered at the group scale. A group, G, is characterized by the number of representatives part of the group, $G(s)$, and the number of patterns attached to the group: $G(t) = \sum_{r \in G} |T(r)|$.

Low density groups are removed. G is a small density group if $G(t) < MinCard$.

High density group representation is enhanced. Let $G(d) = \frac{G(t)}{G(s)}$ be the G group density, and let $\overline{d_g}$ the average density for all the groups. A group G is a high density group if $G(d) > \overline{d_g}$. The number of representatives in G is increased according to the density ratio: $s' = G(s)\left(1 - \frac{G(d)}{\overline{d_g}}\right)$. The s' new representatives in G are randomly chosen.

3.4.1.6 *DIDES*: Illustration of the Whole Process

The main steps of the algorithm are illustrated in Fig. 3.4 with synthetic 2D-data, 40,000 items, structured in clusters of various shapes and densities to which an important level of noise has been added.

The upper part of the figure shows the selected items at two evolution steps. When the proportion P becomes positive (left), the space is not properly covered, some important clusters are not represented, due to the noise. The right picture corresponds to $P > 0.2$. Space coverage is now homogeneous, and the $MaxDmax$

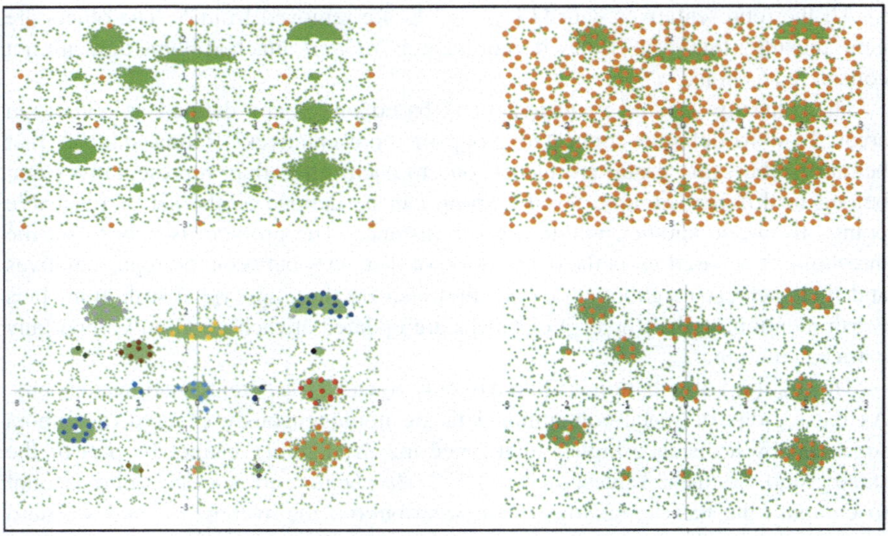

Fig. 3.4 The main steps of the *DIDES* sampling: $P > 0$ (top left), $P > 0.2$ (top right), Output with $g_r = 0.001$ (bottom left) and $g_r = 0.01$ (bottom right)

evolution can be accurately estimated. The fitting curve equation is: $y = 7.1\ x^{0.59}$, with $R^2 = 0.996$.

The lower row of this figure shows the final results for two user granularity settings: $g_r = 0.001$ (left) and $g_r = 0.01$ (right). The granularity is used to compute the $MinCard$ internal parameter which impacts the distance and cardinality thresholds. In both cases, all the noisy representatives have been removed by the density management post-processing and all the clusters are represented. There are more representatives associated to the smaller granularity, even if both cardinalities are similar. In the bottom left graph the connected groups are displayed in distinct colors.

3.4.2 *DENDIS: Density Representation*

While the priority is given to distance in *DIDES*, in *DENDIS* density is of prime concern while distance is controlled: the representative is chosen in the most populated group. A constraint on the hyper volume induced by the samples avoids over sampling in high density areas, thus keeping the sample size small. The attraction basins are not defined using a parameter but are induced by the sampling process.

The stopping criterion is also based upon density: the algorithm stops when there are no more groups with a number of attached patterns higher than the threshold and with a maximum distance in the group high enough with respect to the whole distribution.

The algorithm is made up of two steps. The first one, Algorithm 4, is based on space density while taking into account distance notions. The second one, Algorithm 5, can be seen as a post-processing step which aims at not selecting outliers as representatives.

3.4.2.1 The Algorithm: Choice of the New Representative and Stopping Criterion

This algorithm is called in line 14 of Algorithm 1.

Algorithm 4 *DENDIS*: the selection of the new representative

1: ADD=FALSE, $K = 0.2$, $W_t = n\ g_r$
2: Sort $y_{(1)}, \ldots, y_{(s)}$ with $|T_{y_{(1)}}| \geq \ldots \geq |T_{y_{(s)}}|$
3: **for all** y_k in S **do**
4:     **if** $(|T_{y_k}| < W_t)$ **then**
5:         break
6:     **end if**
7:     $\alpha_k = max(\frac{W_t}{|T_{y_k}|}, K)$
8:     **if** $(d_{max}(y_k) \geq \alpha_k\ \overline{\underset{y_k \in S}{d_{max}(y_k)}})$ **then**
9:         $y_* = x_{max}(y_k)$
10:         ADD=TRUE, break
11:     **end if**
12: **end for**

The already selected items, y, are sorted according to the cardinality of the set of patterns they are the representative (line 2) and these sets, T_{y_k}, are analyzed in decreasing order of weight. Each of them is split when two conditions are met (lines 4 and 8). The first one deals with the number of attached patterns: it has to be higher than the threshold, $W_t = n\ g_r$. Without any additional constraint, the representatives would tend to have the same number of patters attached, close to W_t. This behavior would lead to an oversized sample in high density areas. Therefore, the other condition is related to the density, controlled by the induced hyper volume. At the beginning of the process, the d_{max} values are quite high, as well as the cardinalities $|T_{y_k}|$. The fraction of minimum volume is then limited by an upper bound, K. This allows for the space to be covered in an homogeneous way, the d_{max} values tend to a lower mean with a lower deviation. In the last steps of the process, α_k dynamically promotes dense areas in order for the sample to reflect the original densities: the larger the cardinality the smaller α_k and thus the constraint on the induced volume (line 7). The constant value, $K = 0.2$, has been empirically defined from experimental simulations.

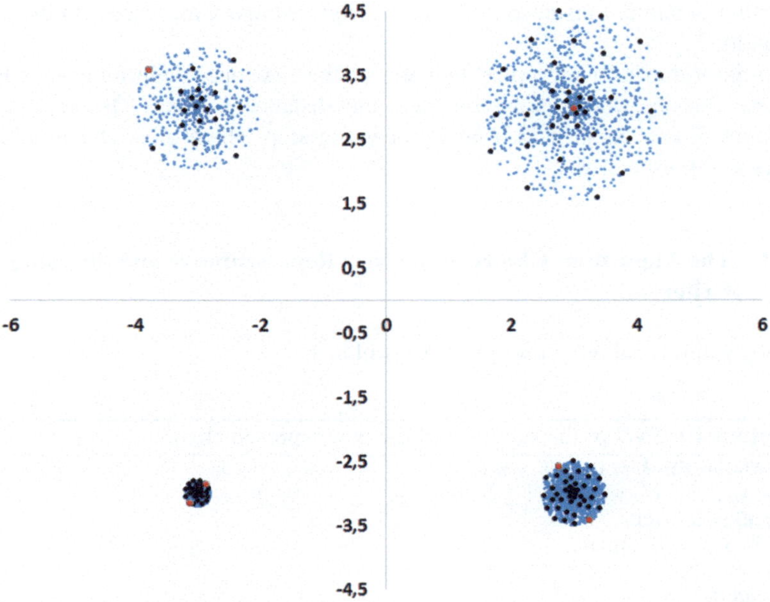

Fig. 3.5 Impact of the induced volume constraint

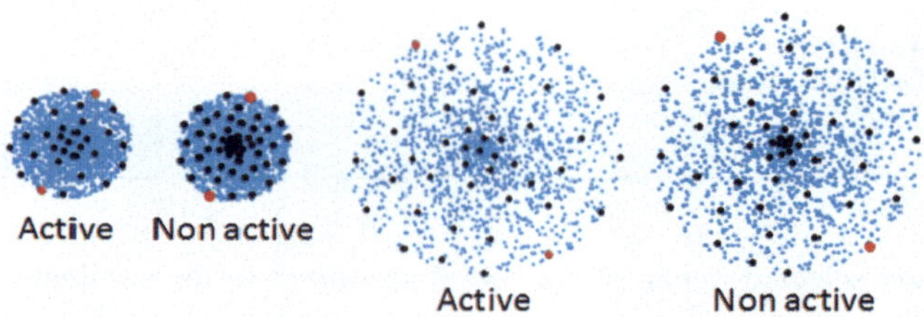

Fig. 3.6 Zoom of the two densest clusters

Figure 3.5 illustrates the impact of the constraint on the induced volume (Algorithm 4, line 5). The data (blue) are well structured in four clusters of heterogeneous densities. The six first selected representatives are plotted in red, while the following ones appear in black. The small groups, in the bottom part of the figure, are denser than the others. The results for these two clusters are displayed in a zoom version in Fig. 3.6, with and without this constraint.

Without the mentioned constraint, the new representatives are located in the denser areas until the number of attached patterns become smaller than the W_t threshold. When the constraint is active the number of representatives in the dense area is limited by the induced volume. Density and distance are both useful to avoid an over representation in dense areas.

As previously explained, the new representative is chosen as the furthest attached pattern, $x_{max}(y_k)$, for space covering purposes (line 9).

The process is repeated until there is no more set to split (line 4–5).

3.4.2.2 The Post-Processing Step

When all the s representatives are selected, the post-processing step, Algorithm 5, discards outliers as representatives. As the new selected item is chosen as the furthest from the ones which are already selected, the S set is likely to include some outliers.

Algorithm 5 *DENDIS*: the post-processing algorithm

1: **for all** y_i in S **do**
2:     **if** $(|T_{y_i}| \leq W_n)$ **then**
3:         $S = S - \{y_i\}$, $s = s - 1$
4:     **end if**
5:     **if** $(d_{max}(y_i) > \overline{d_{max}(y_i)}_{y_i \in S})$ **then**
6:         $y_i = \underset{x_l \in T_{y_i}}{arg\ min}\ d(x_l, B)$ { B is the barycenter of T_{y_i} }
7:     **end if**
8:     $T_{y_i} = \{y_i\}$
9: **end for**
10: **for all** $x_l \in T \setminus S$ **do**
11:     Find $d_{near}(x_l) = \min_{y_k \in S} d(x_l, y_k)$
12:     $T_{y_k} = T_{y_k} \cup \{x_l\}$
13: **end for**

Two cases may occur.

In the first one (lines 2–3), when the representative is isolated, the number of attached patterns is lower than or equal to the noise threshold $|T_{y_i}| \leq W_n$, inferred from the T_{y_k} distribution. Let $T' = \{T_{y_k} |\ |T_{y_k}| < \overline{|T_{y_k}|}\}$ be the reduced set of representatives with a number of attached pattern less than the average, and let m, σ, and min, the mean, standard deviation, and minimum of the $|T'|$. The noise threshold is defined as: $W_n = max(m - 2\sigma, min)$. The choice is then to remove this representative labeled as noise.

In the other case, the outlier detection is based upon the induced volume: the corresponding d_{max} is higher than average (line 5). In the post-processing phase, the input space coverage is quite homogeneous: the mean can be used as a threshold. In this case, the new representative is chosen as the closest to the barycenter, B, of the set (line 6). This way of doing is similar to the usual practice: the representative is set at the center of the dense areas, like in kernel and neighboring approaches. By contrast, the proposal comes to select the representative at the border of the dense area. Once at least a representative has been changed or removed, an update of the attached patterns is needed (lines 10–13), and to do this the sets of attached patterns must be previously reset (line 8).

Even if *DIDES* and *DENDIS* look similar, with the same unique parameter and a common reference to distance and density, they are very different in the way these concepts are combined. *DENDIS* is able to deal with highly structured data while *DIDES* is more suited to fit irregular shapes.

3.4.3 *ProTraS: Based on the Sampling Cost*

DIDES and *DENDIS* are powerful algorithms that manage in different ways distance and density using in single user parameter, $granularity$. However, they also have limitations. First, the meaning that this unique parameter conveys is not the same for the two proposals, this may be an inconvenient from the user point of view. Second, they use internal heuristics based on parameters inferred from the data. Although the approach has been validated on a wide range of data sets with contrasted characteristics, heuristics elimination is always advisable.

The *fft*-based algorithms yield coresets even if they were not designed with this goal in mind. The objective of *ProTras* is to generate coresets.

3.4.3.1 Description of the Algorithm

The proposed sampling algorithm generates a (k, ε) coreset, S, for a whole set T. The sampling cost plays a central role. The approximation level, ε, is the unique parameter. It also serves as the stopping criterion: the algorithm ends when the cost is below the threshold. Finally it is used to guide the sampling process itself: at each iteration a new representative is added to the sample, in the group with the highest probability of cost reduction. This tends to limit the sample size.

The algorithm is summarized in Fig. 3.7.

The representative is the farthest-first traversal item of a given group.

As the sampling cost in a given group, j, is proportional to both the number of attached patterns, $w_j = |T_{y_j}|$, and the maximum within group distance, $d_j = d_{max}(y_j)$, the new representative is chosen in the group that is likely to best contribute to the global cost, ε, reduction. The probability of cost reduction is a combination of two basic probabilities: the first one according to the distance and the

Fig. 3.7 *ProTraS*: summary of the algorithm

1. Add a new sample in the group with the highest probability of cost reduction
2. Assign each pattern to the nearest sample
3. Compute Cost
4. If (Cost > CostParam) goto Step 1

other one according to the number of attached patterns. Each of them is estimated by normalizing the values by the maximum over all the groups.

For a given group, j, the probabilities are:

- Density-based probability: $P_{dens}(j) = \frac{w_j}{\max\limits_i w_i}$
- Distance-based probability: $P_{dist}(j) = \frac{d_j}{\max\limits_i d_i}$

Both have the same meaning: the higher the value, the higher the expected cost reduction. The probability of cost reduction is computed as the following combination: $P_{CostRed}(j) = \frac{w_j d_j}{\max\limits_i (w_i d_i)}$.

This probabilistic approach differs from greedy algorithms. For instance in decision trees, the node to be split is the one with the highest gain. This requires all the nodes to be tested, i.e., all the gains have to be computed. The probability estimation tends to speed up the sampling process.

The algorithm is detailed in Algorithm 6.

Algorithm 6 The *ProTraS* algorithm

1: Input: $T = \{x_i\}, i = 1 \ldots, n, \varepsilon$
2: Output: $S = \{y_j\}, T(y_j), j = 1, \ldots, s$
3: Select an initial pattern $x_{init} \in T$
4: $S = \{y_1 = x_{init}\}, s = 1, T(y_1) = \{y_1\}$
5: **repeat**
6:     **for all** $x_l \in T \setminus S$ **do**
7:         Find $d_{near}(x_l) = \min\limits_{y_k \in S} d(x_l, y_k)$
8:         $T(y_k) = T(y_k) \cup \{x_l\}$ {Set of patterns represented by y_k}
9:     **end for**
10:     $MaxWD = 0, Cost = 0$
11:     **for all** $y_k \in S$ **do**
12:         Find $d_{max}(y_k) = \max\limits_{x_m \in T(y_k)} d(x_m, y_k)$
13:         Store $d_{max}(y_k), x_{max}(y_k)$ {where $d_{max}(y_k) = d(x_{max}(y_k), y_k)$}
14:         $p_k = |T(y_k)|\, d_{max}(y_k)$
15:         **if** $(p_k > MaxWD)$ **then**
16:             $MaxWD = p_k, y_w = y_k$
17:         **end if**
18:         $Cost := Cost + p_k/n$
19:     **end for**
20:     $y_* = x_{max}(y_w)$
21:     $S = S \cup \{y_*\}, s = s + 1, T(y_s) = \{y_*\}$
22:     Update Cost
        {Remove the part due to former y_* and add the two new costs, induced by y_w and y_*}
23: **until** $(Cost < \varepsilon)$
24: **return** $Cost, S, T(y_j), j = 1, \ldots, s$

The main loop, lines 5–23, includes two *for* loops. The first one is described in Algorithm 1. The second loop is enriched compared to the previous version. It also computes the combined probability according to $|T_{y_k}|$ and $d_{max}(y_k)$ (line 14), identifies the group with the highest cost reduction probability (line 16), and updates the global cost before splitting (line 18). The new representative is the farthest-first traversal, y_*, of the group with the highest probability, represented by y_w. To update the current cost, the cost due to former y_* is subtracted and the new costs, induced by y_w and y_*, are added. The cost is computed at each iteration without any additional computational effort.

The algorithm uses, in line 18, a normalized distortion cost: $\frac{\sum_{j=1}^{s} w_j\ d_j}{n}$, meaning that the approximation, ε, is now a proportion of the whole normalized cost, $\frac{Cost_T(C)}{n}$. This new expression carries the same meaning for all the data, irrespective of the data size.

This last algorithm is really simple as it is only driven by the sampling cost. The stopping criterion is easy to understand and no post-processing is required contrary to *DIDES* and *DENDIS*.

3.5 Algorithms' Behavior: Common Properties and Some Differences

As they are based on similar concepts, the three algorithms in the family share common properties but they also have their own characteristics. The former and the latter are studied in this section.

3.5.1 Common Properties

Some interesting properties shared by the three algorithms are illustrated using the $S1$ data shown in Fig. 3.8.

This synthetic data set includes 3000 $2D$-points and was used in [30] under the name *A.set 1*.

3.5.1.1 Little Sensitivity to Initialization

The first sample item is randomly chosen. This section aims to assess how this unique random step impacts the sampling process.

This experiment is based upon the distance between a sample point in a given *ProTraS* outcome and its nearest neighbor in another one. This distance is expected to be low on average if the sampling process is only slightly impacted by the initialization step.

Let S_1 and S_2 be two sample sets given by two runs with the same cost, and s_1, s_2 their corresponding sizes. For a given $x \in S_1$, the distance with its closest neighbor,

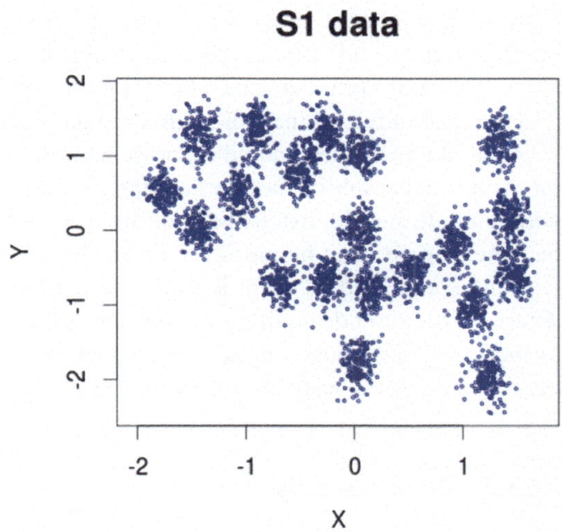

Fig. 3.8 The $S1$ data

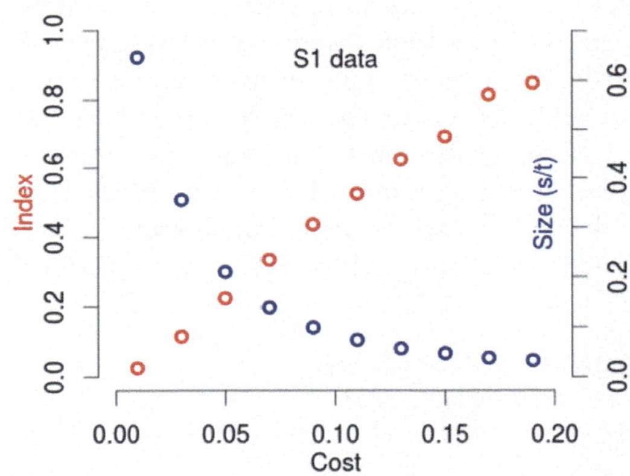

Fig. 3.9 Sensitivity to initialization, index normalization with d for 1% of the data

$y \in S_2$ is $d_{nn}(x, y)$. The index used for the two sample set comparison is computed as follows:

$$I(S_1, S_2) = \frac{1}{s_1 s_2 d} \left(\sum_{x \in S_1} d_{nn}(x, y) + \sum_{y \in S_2} d_{nn}(y, x) \right)$$

where d is a normalization distance. This distance could be chosen as the average distance between all points in both sets. But, in this case, the ratio would be quite low and difficult to interpret. Instead, d is the mean radius of the hypersphere that includes a given proportion of the data.

This index is averaged over 100 trials for a given cost. The results for $S1$ and d computed for 1% of the data are plotted in Fig. 3.9.

A wide range of costs was studied. This experiment confirms that random initialization has no critical influence on the results: even with small sample sets, till $s/t = 1\%$, the average distance between the neighbors of two different outcomes is less than the radius of the sphere that includes 1% of the data.

As all the instances, including noise and outliers, are given the same importance, this empirical result cannot be guaranteed for all the configurations. However, it is widely confirmed by extensive experiments. In the most extreme cases, this radius becomes the one that includes 5% of the data.

This is expected as there is only one random step. Making the algorithm fully deterministic would require an extra iteration to select the first sample item as the farthest from a virtual extreme pattern computed as the minimum (or the maximum) over the data set in each dimension.

3.5.1.2 Robust to Noise

An increasing amount of uniform random noise (from 1 to 9%) was added to the data. The new values were computed independently in each dimension, according to the whole range of the given feature: $noise_j = \min_j + U[0.+1] * (\max_j - \min_j)$.

Table 3.1 reports the sample size and the RI.

Noise clearly impacts the sampling: the sample size tends to increase with the amount of noise, even if these results also include an initialization variation. RI values show that the sampling algorithm is still able to identify and represent the data structure even with a significant level of noise.

Table 3.1 Noise sensitivity: sample size and RI, on $S1$ data, $cost = 0.1$

% noise	Size	s	RI
0	3000	261	0.981
1	3030	272	0.972
2	3060	278	0.973
3	3090	300	0.986
4	3120	301	0.990
5	3150	317	0.980
6	3180	317	0.990
7	3210	326	0.982
8	3240	325	0.985
9	3270	333	0.963
μ		303	0.980
σ		25.1	0.009

3.5.1.3 Sample Size Mainly Dependent on Data Structure

The test consists in comparing the algorithm performances on the initial set and an enriched similar one. The latter results from the aggregation of new patterns: each data point was replicated k times, $k = 2, \ldots 20$, with an additional uniform random noise in the range $[-0.1, +0.1]\sigma_j$ where σ_j is the standard deviation for feature j. It is worth mentioning that the noise generation is distinct from the one used in the previous section. The expected magnitude is lower as the goal is just to generate non identical items. The sampling is applied with the same cost. In Table 3.2 the sample size and the RI are given for the initial set and the enriched ones.

The sample size is significantly increased at the first step of the procedure due to the additional noise: the data structure is significantly modified as the noise is randomly defined for each input dimension. Then it tends to become more stable. The sample size ranges from 261 to 397 while the whole size becomes 20 times bigger. This does not impact the results. Moreover, for the higher values of k, between 30,000 and 60,000, the variation, from 371 to 397, is due to the initialization of the *k-means* algorithm: the same results are obtained when the duplication is done without additional noise. This experiment shows that the outcome of *ProTraS* depends more on the data structure than on the data size.

Table 3.2 Data size sensitivity: sample size and RI, on $S1$ data, $cost = 0.1$

Size	s	RI
3000	261	0.987
6000	316	0.973
9000	331	0.980
12,000	338	0.977
15,000	347	0.983
18,000	354	0.970
21,000	363	0.980
24,000	357	0.981
27,000	366	0.979
30,000	371	0.980
33,000	368	0.983
36,000	377	0.991
39,000	374	0.981
42,000	386	0.980
45,000	376	0.985
48,000	382	0.985
51,000	388	0.986
54,000	394	0.975
57,000	397	0.977
60,000	379	0.986
μ		0.981
σ		0.005

Due to the similar behavior of the two algorithms, the comparison with alternative sampling techniques carried out with *DENDIS* in [52] holds for *ProTraS*.

3.5.2 Some Differences

To investigate the algorithms' behavior two synthetic data sets were designed. They are spatially structured in clusters of various shapes and densities. The first one, $SD1$, is made up of 3800 $2D$-points. The second one, $SD2$, is based on data widely used by the scientific community [31], and includes 8000 $2D$-points. The input density is quite homogeneous.

These data can be processed by the two algorithms with the same unique parameter, granularity. The final distortion cost can be computed in both cases. *ProTraS* can also be run with different cost values. The comparison between these three algorithms can be based on the sample size, the sampling cost, and also on the sample item locations.

As an example, the granularity $g_r = 0.007$ is considered for the $SD1$ data. The results obtained with *DIDES* and *DENDIS* are very different. The former yields a sample set of 154 items and the corresponding cost is 0.122. These values become 224 and 0.086 with the latter. This is quite interesting: the relationship between cost and granularity is not the same for the two algorithms. This is also expected because granularity is combined with the data size to define a threshold, but this threshold is not used in the same way in both cases, as recalled in the heading of Sect. 3.4.

A fair comparison cannot be based upon the granularity value: either the sample size or the cost should be similar. The granularity value that gives with *DENDIS* results close to those of *DIDES* with $g_r = 0.007$ is $g_r = 0.013$: 123 representatives and the same cost of 0.122.

When *ProTraS* is run with this cost, the sample size is 128.

Figure 3.10 shows that the three algorithms behave in a comparable but distinct way. *DIDES* ensures a space coverage without taking density into account. The sample points are spatially distributed to fit data shapes. Using *DENDIS* the representatives are located in high density areas. *ProTraS*, using the combined probabilities, gives a trade-off between space coverage and density representation. In high density areas, *ProTraS* uses fewer sample points than *DENDIS* but more than *DIDES*. In more sparsely populated areas, it is the opposite.

For the $SD1$ data, with the same number of representatives, *DENDIS* yields a lower cost than *DIDES*. This can be explained by the difference in their densities. More representatives in high density zones decrease the global cost.

As the $SD1$ data set includes four connected subsets, a smart algorithm should identify the same number of clusters. $DBSCAN$ [17] was run on the whole data set as well as on the $ProTraS$ sample with a 0.1 cost. Figure 3.11 shows the silhouette index [57] evolution according to the distance, d, that defines the neighborhood for directly reachable points.

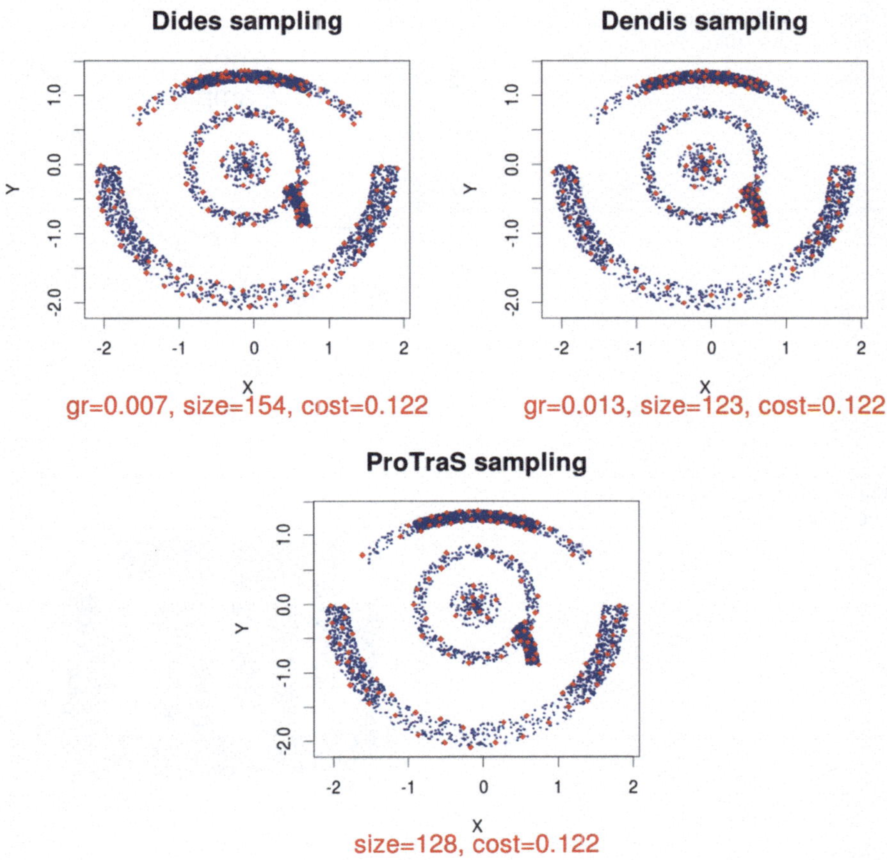

Fig. 3.10 The three sampling algorithms with $SD1$ data

In both cases, either from the whole data set or from the sample, the optimal number of clusters is clearly identified.

To get a global picture, these two algorithms were run for granularity values ranging from 0.001 to 0.02, with 100 steps for the two data sets $SD1$ and $SD2$. Density is more homogeneous in the $SD2$ data, as shown in Fig. 3.12. Both the cost and the sample size were recorded. The cost evolution is shown in Fig. 3.13.

The difference in the cost values depends on the algorithm, with *DIDES* yielding higher costs than *DENDIS*, but also on the data. With $SD1$, the gap between the two curves increases with the granularity. When the density is more homogeneous, the two sampling methods give more similar results as shown in the right part of Fig. 3.13.

This can be explained by the evolution of the sample size, shown in Fig. 3.14.

Fig. 3.11 Silhouette index evolution with *DBSCAN* on *SD*1 data

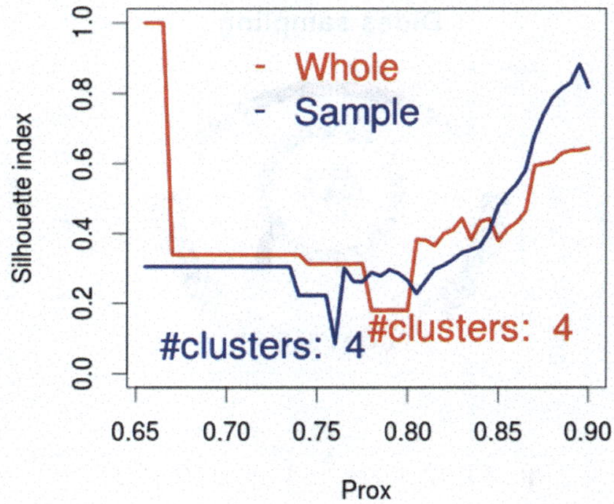

Fig. 3.12 *ProTraS* sampling with *SD*2 data

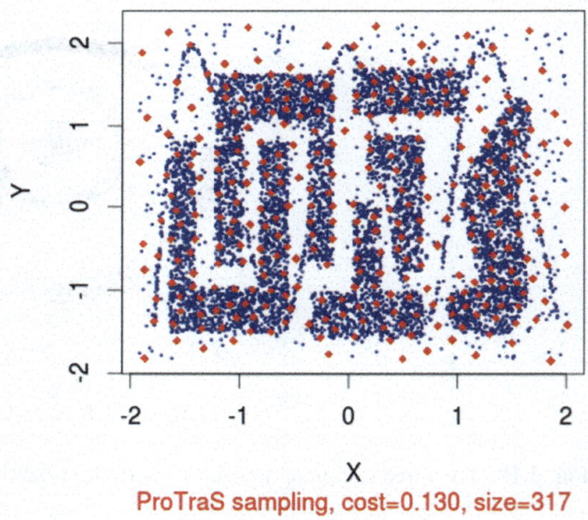

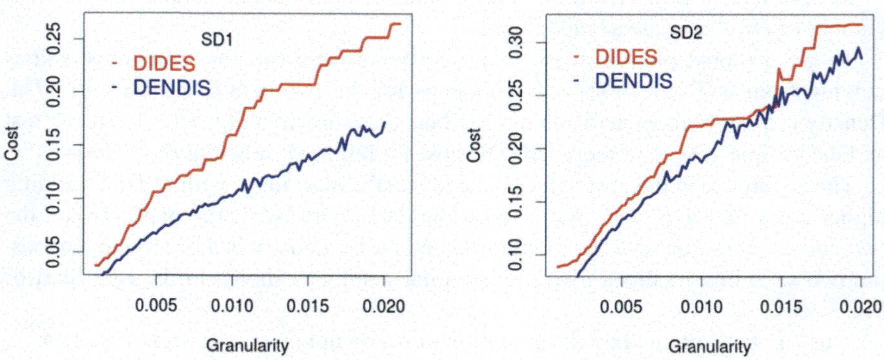

Fig. 3.13 Cost vs granularity for *SD*1 and *SD*2 data

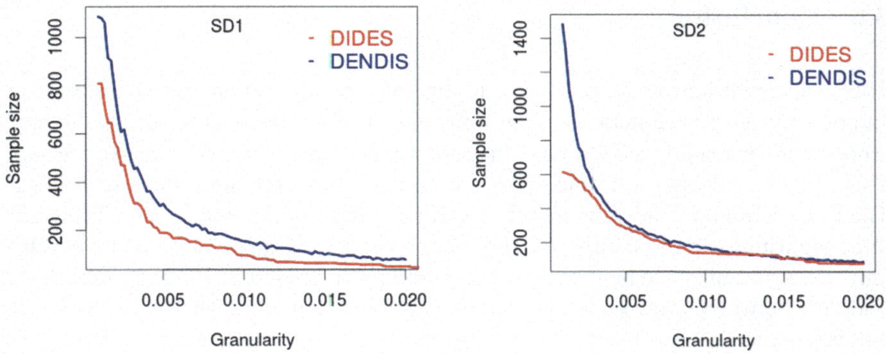

Fig. 3.14 Size vs granularity for $SD1$ and $SD2$ data

Fig. 3.15 Cost vs Sample size for the three algorithms with $SD1$ and $SD2$ data

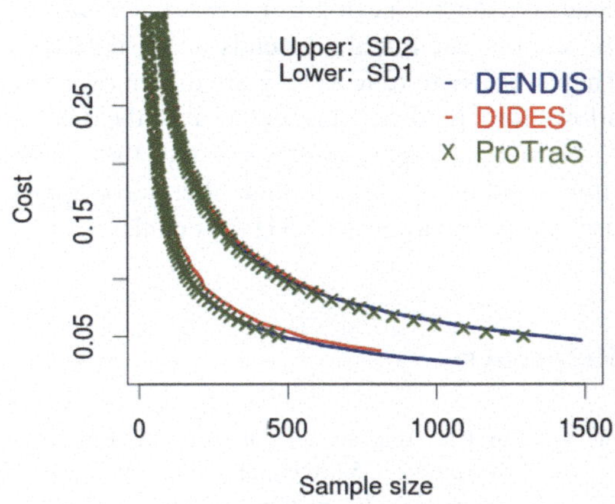

The *DIDES* sample size is lower than the *DENDIS* one for a given granularity. The difference is greater with $SD1$ data. This seems to be in agreement with a higher cost.

Even if granularity has a different meaning for each algorithm, the relationship between the sample size and the cost is almost monotone, and the same for the three algorithms as shown in Fig. 3.15. This result is expected as the three of them are *fft*-like.

Some differences can be observed. *DIDES* is the least efficient in terms of the cost size ratio, while *ProTraS* and *DENDIS* have a similar behavior. It is worth mentioning that this comparable, or even better, result is gained with a very simple algorithm, which specifically aims at a given cost approximation. No other parameters, explicit or hidden, are needed.

3.6 Conclusion

This chapter introduces a family of sampling algorithms that are based on distance and density. These are iterative algorithms that share three concepts. First, the new representative at a given iteration is the farthest-first traversal item in a given group. This property allows for a time optimization that makes the algorithms really fast. The third common concept is that they yield a coreset of the whole data. There are three algorithms in the family. *DIDES* tends to cover the input space and *DENDIS* aims to represent space densities, the probabilistic approach in *ProTraS* achieves a trade-off between space coverage and density representation. The latter explicitly builds a coreset.

The three of them have common properties. They show little sensitivity to initialization as only the first sample item is randomly chosen and can be made fully deterministic by starting from a virtual extreme point. They are robust to noise and the sample size mainly depends on the data structure instead of the data size itself. They are easy to tune as they are driven by a unique and meaningful parameter that allows for a gradual representation of the data.

Numerical experiments show that they are really faster than competitive algorithms and more efficient than alternative approaches. The reader may refer to the original publications [52–54] for details.

References

1. Agarwal, P.K., Har-Peled, S., Varadarajan, K.R.: Approximating extent measures of points. J. ACM **51**(4), 606–635 (2004). http://doi.acm.org/10.1145/1008731.1008736
2. Al-Kateb, M., Lee, B.: Adaptive stratified reservoir sampling over heterogeneous data streams. Inf. Syst. **39**(1), 199–216 (2014)
3. Al-Kateb, M., Lee, B.S., Wang, X.S.: Adaptive-size reservoir sampling over data streams. In: 19th International Conference on Scientific and Statistical Database Management, SSBDM'07, pp. 22–22. IEEE, Piscataway (2007)
4. Arthur, D., Vassilvitskii, S.: k-means++: the advantages of careful seeding. In: Proceedings of the Eighteenth Annual ACM-SIAM Symposium on Discrete Algorithms, pp. 1027–1035. Society for Industrial and Applied Mathematics, Philadelphia (2007)
5. Azzalini, A., Torelli, N.: Clustering via nonparametric density estimation. Stat. Comput. **17**(1), 71–80 (2007)
6. Bezdek, J.C.: Pattern Recognition with Fuzzy Objective Functions Algorithms. Plenum Press, New York (1981)
7. Celebi, M.E., Kingravi, H.A., Vela, P.A.: A comparative study of efficient initialization methods for the k-means clustering algorithm. Expert Syst. Appl. **40**(1), 200–210 (2013)
8. Chaudhuri, S., Das, G., Narasayya, V.: Optimized stratified sampling for approximate query processing. ACM Trans. Database Syst. **32**(2), 9 (2007)
9. Chehreghani, M., Abolhassani, H., Chehreghani, M.: Improving density-based methods for hierarchical clustering of web pages. Data Knowl. Eng. **67**(1), 30–50 (2008)
10. Chernoff, H.: A measure of asymptotic efficiency for tests of a hypothesis based on the sum of observations. Ann. Math. Stat. **23**(4), 493–507 (1952)

11. Chernoff, H.: A note on an inequality involving the normal distribution. Ann. Probab. **9**, 533–535 (1981)
12. Chiang, M.C., Tsai, C.W., Yang, C.S.: A time-efficient pattern reduction algorithm for k-means clustering. Inf. Sci. **181**(4), 716–731 (2011)
13. Chiu, S.L.: Fuzzy model identification based on cluster estimation. J. Intell. Fuzzy Syst. **2**, 267–278 (1994)
14. Dolnicar, S., Leisch, F.: Segmenting markets by bagged clustering. Australas. Mark. J. **12**(1), 51–65 (2004)
15. Efraimidis, P.S., Spirakis, P.G.: Weighted random sampling with a reservoir. Inf. Process. Lett. **97**(5), 181–185 (2006)
16. Epanechnikov, V.A.: Non-parametric estimation of a multivariate probability density. Theory Probab. Appl. **14**(1), 153–158 (1969)
17. Ester, M., Kriegel, H.P., Sander, J., Xu, X.: A density-based algorithm for discovering clusters in large spatial databases with noise. In: Proceedings of the 2nd International Conference on Knowledge Discovery and Data Mining, pp. 226–231. AAAI Press, Menlo Park (1996)
18. Feldman, D., Faulkner, M., Krause, A.: Scalable training of mixture models via coresets. In: Advances in Neural Information Processing Systems, pp. 2142–2150 (2011)
19. Gutmann, B., Kersting, K.: Stratified gradient boosting for fast training of conditional random fields. In: Proceedings of the 6th International Workshop on Multi-Relational Data Mining, pp. 56–68 (2007)
20. Har-Peled, S., Mazumdar, S.: On coresets for k-means and k-median clustering. In: Proceedings of the Thirty-Sixth Annual ACM Symposium on Theory of Computing, STOC'04, pp. 291–300. ACM, New York (2004). https://doi.org/10.1145/1007352.1007400
21. Hartigan, J.A.: Clustering Algorithms. Wiley, London (1975)
22. Hartigan, J.A., Wong, M.: A k-means clustering algorithm. Appl. Stat. **28**, 100–108 (1979)
23. Hatamlou, A., Abdullah, S., Nezamabadi-pour, H.: A combined approach for clustering based on k-means and gravitational search algorithms. Swarm Evol. Comput. **6**, 47–52 (2012)
24. Hochbaum, D.S., Shmoys, D.B.: A best possible heuristic for the k-center problem. Math. Oper. Res. **10**(2), 180–184 (1985)
25. Hodge, V.J., Austin, J.: A survey of outlier detection methodologies. Artif. Intell. Rev. **22**(2), 85–126 (2004)
26. Hoeffding, W.: Probability inequalities for sums of bounded random variables. J. Am. Stat. Assoc. **58**(301), 13–30 (1963)
27. Ilango, M.R., Mohan, V.: A survey of grid based clustering algorithms. Int. J. Eng. Sci. Technol. **2**(8), 3441–3446 (2010)
28. Jain, A.K.: Data clustering: 50 years beyond k-means. Pattern Recognit. Lett. **31**(8), 651–666 (2010)
29. Jiang, M.F., Tseng, S.S., Su, C.M.: Two-phase clustering process for outliers detection. Pattern Recognit. Lett. **22**(6), 691–700 (2001)
30. Kärkkäinen, I., Fränti, P.: Dynamic local search algorithm for the clustering problem. Tech. Rep. A-2002-6, Department of Computer Science, University of Joensuu, Joensuu (2002)
31. Karypis, G., Han, E.H., Kumar, V.: Chameleon: hierarchical clustering using dynamic modeling. Computer **32**(8), 68–75 (1999)
32. Kaufman, L., Rousseeuw, P.: Clustering by means of medoids. In: Statistical Data Analysis Based on the L1-Norm and Related Methods. North-Holland, Amsterdam (1987)
33. Kerdprasop, K., Kerdprasop, N., Sattayatham, P.: Density-biased clustering based on reservoir sampling. In: Proceedings Sixteenth International Workshop on Database and Expert Systems Applications, pp. 1122–1126. IEEE, Piscataway (2005)
34. Khan, S.S., Ahmad, A.: Cluster center initialization algorithm for k-modes clustering. Expert Syst. Appl. **40**(18), 7444–7456 (2013)
35. Kollios, G., Gunopulos, D., Koudas, N., Berchtold, S.: Efficient biased sampling for approximate clustering and outlier detection in large data sets. IEEE Trans. Knowl. Data Eng. **15**(5), 1170–1187 (2003)

36. Krishnapuram, R., Keller, J.M.: A possibilistic approach to clustering. IEEE Trans. Fuzzy Syst. **(1)**, 98–110 (1993)
37. Leisch, F., Dolnicar, S.: Winter tourist segments in austria: identifying stable vacation styles using bagged clustering techniques. J. Travel Res. **41**(3), 281–292 (2003)
38. Linde, Y., Buzo, A., Gray, R.: An algorithm for vector quantizer design. IEEE Trans. Commun. **28**(1), 84–95 (1980). https://doi.org/10.1109/TCOM.1980.1094577
39. Ling, R.F.: Cluster analysis algorithms for data reduction and classification of objects. Technometrics **23**(4), 417–418 (1981)
40. Lloyd, S.P.: Least squares quantization in PCM. IEEE Trans. Inf. Theory **28**(2), 129–137 (1982)
41. Lv, Y., Ma, T., Tang, M., Cao, J., Tian, Y., Al-Dhelaan, A., Al-Rodhaan, M.: An efficient and scalable density-based clustering algorithm for datasets with complex structures. Neurocomputing **171**, 9–22 (2015)
42. Ma, X., Pan, Z., Li, Y., Fang, J.: High-quality initial codebook design method of vector quantisation using grouping strategy. IET Image Process. **9**, 986–992 (2015)
43. Machová, K., Puszta, M., Barčák, F., Bednár, P.: A comparison of the bagging and the boosting methods using the decision trees classifiers. Comput. Sci. Inf. Syst. **3**(2), 57–72 (2006)
44. Macqueen, J.: Some methods for classification and analysis of multivariate observations. In: In 5-th Berkeley Symposium on Mathematical Statistics and Probability, pp. 281–297 (1967)
45. Menardi, G., Azzalini, A.: An advancement in clustering via nonparametric density estimation. Stat. Comput. **24**(5), 753–767 (2014)
46. Mitra, P., Murthy, C., Pal, S.: Density-based multiscale data condensation. IEEE Trans. Pattern Anal. Mach. Intell. **24**(6), 734–747 (2002)
47. Naldi, M., Campello, R.: Comparison of distributed evolutionary k-means clustering algorithms. Neurocomputing **163**, 78–93 (2015)
48. Nanopoulos, A., Manolopoulos, Y., Theodoridis, Y.: An efficient and effective algorithm for density biased sampling. In: Proceedings of the Eleventh International Conference on Information and knowledge Management, pp. 398–404 (2002)
49. Nanopoulos, A., Theodoridis, Y., Manolopoulos, Y.: Indexed-based density biased sampling for clustering applications. Data Knowl. Eng. **57**(1), 37–63 (2006)
50. Palmer, C.R., Faloutsos, C.: Density biased sampling: an improved method for data mining and clustering. In: ACM SIGMOD Intl. Conference on Management of Data, Dallas, pp. 82–92 (2000)
51. Rahman, M.A., Islam, M.Z.: A hybrid clustering technique combining a novel genetic algorithm with k-means. Knowl.-Based Syst. **71**, 345–365 (2014)
52. Ros, F., Guillaume, S.: Dendis: A new density-based sampling for clustering algorithm. Expert Syst. Appl. **56**, 349–359 (2016). https://doi.org/10.1016/j.eswa.2016.03.008
53. Ros, F., Guillaume, S.: Dides: a fast and effective sampling for clustering algorithm. Knowl. Inf. Syst. **50**, 543–568 (2016). https://doi.org/10.1007/s10115-016-0946-8
54. Ros, F., Guillaume, S.: Protras: a probabilistic traversing sampling algorithm. Expert Syst. Appl. **105**, 65–76 (2018). https://doi.org/10.1016/j.eswa.2018.03.052
55. Ros, F., Pintore, M., Deman, A., Chrétien, J.: Automatical initialization of RBF neural networks. Chemom. Intell. Lab. Syst. **87**(1), 26–32 (2007)
56. Rosenkrantz, D.J., Stearns, R.E., Lewis, P.M. II.: An analysis of several heuristics for the traveling salesman problem. SIAM J. Comput. **6**(3), 563–581 (1977)
57. Rousseeuw, P.J.: Silhouettes: a graphical aid to the interpretation and validation of cluster analysis. J. Comput. Appl. Math. **20**, 53–65 (1987)
58. Sarma, T., Viswanath, P., Reddy, B.: Speeding-up the kernel k-means clustering method: a prototype based hybrid approach. Pattern Recognit. Lett. **34**(5), 564–573 (2013)
59. Sarma, T.H., Viswanath, P., Reddy, B.E.: Speeding-up the kernel k-means clustering method: a prototype based hybrid approach. Pattern Recognit. Lett. **34**(5), 564–573 (2013)
60. Tan, S.C., Ting, K.M., Teng, S.W.: A general stochastic clustering method for automatic cluster discovery. Pattern Recognit. **44**(10), 2786–2799 (2011)
61. Tax, D., Duin, R.: Support vector data description. Mach. Learn. **54**(1), 45–66 (2004)

62. Viswanath, P., Sarma, T., Reddy, B.: A hybrid approach to speed-up the k-means clustering method. Int. J. Mach. Learn. Cybern. **4**(2), 107–117 (2013)
63. Vitter, J.S.: Random sampling with a reservoir. ACM Trans. Math. Softw. **11**(1), 37–57 (1985)
64. Wang, X., Wang, X., Wilkes, D.M.: A divide-and-conquer approach for minimum spanning tree-based clustering. IEEE Trans. Knowl. Data Eng. **21**(7), 945–958 (2009)
65. Xiao, Y., Liu, B., Hao, Z., Cao, L.: A k-farthest-neighbor-based approach for support vector data description. Appl. Intell. **41**(1), 196–211 (2014)
66. Yager, R.R., Filev, D.P.: Generation of fuzzy rules by mountain clustering. J. Intell. Fuzzy Syst. **2**, 209–219 (1994)
67. Yang, M.S., Wu, K.L.: A modified mountain clustering algorithm. Pattern Anal. Appl. **8**(1–2), 125–138 (2005)
68. Zahra, S., Ghazanfar, M.A., Khalid, A., Azam, M.A., Naeem, U., Prugel-Bennett, A.: Novel centroid selection approaches for kmeans-clustering based recommender systems. Inf. Sci. **320**, 156–189 (2015)
69. Zhong, C., Malinen, M., Miao, D., Fränti, P.: A fast minimum spanning tree algorithm based on k-means. Inf. Sci. **295**, 1–17 (2015)

Chapter 4
From Supervised Instance and Feature Selection Algorithms to Dual Selection: A Review

Frédéric Ros and Serge Guillaume

4.1 Introduction

For classification problems there is a target concept we wish to learn about and this concept is represented in each instance as a dependent variable having finitely many values called the class. The goal of classification is therefore to construct a model from instances with known class values, called the training dataset, to predict which class an instance should belong to in view of its observed attribute values. In some cases the induced classification model, or classifier, is simply used to make class predictions for instances with unknown class values. In supervised learning, a training set providing previously known information is used to classify new instances. Commonly, several instances are stored in the training set but some of them are not useful for classification purposes; therefore, it is possible to obtain acceptable classification rates while ignoring non-useful cases; this process is known as instance selection. Through instance selection [13, 77] the training set is reduced, which also reduces runtimes in the classification and/or training stages of classifiers.

Over recent decades, database sizes have grown considerably. Big data is characterized by an increase in the volume, velocity, and variety of the data. From a technical point of view, big data are required for cleaning, analyzing, securing, and providing a granular access to massive datasets. Big data present new challenges, because machine learning algorithms were not designed to process such large

F. Ros (✉)
PRISME Laboratory, Orléans University, Orléans, France
e-mail: frederic.ros@univ-orleans.fr

S. Guillaume
UMR ITAP, Irstea, Montpellier, France
e-mail: serge.guillaume@irstea.fr

F. Ros, S. Guillaume (eds.), *Sampling Techniques for Supervised or Unsupervised Tasks*, Unsupervised and Semi-Supervised Learning,
https://doi.org/10.1007/978-3-030-29349-9_4

volumes of information. The application of classical algorithms is much more difficult and scalability is a major concern. With regard to instance selection, the solutions that have been proposed so far to deal with big data problems adopt the divide and conquer approach. Instance selection methods can alleviate this problem when the size of the dataset is medium to large. However, even these methods face similar problems with very large-to-massive datasets.

In addition to larger sizes, big data usually conducted by various fields of science may contain hundreds of features. As most of these features have no relevance with data mining tasks, they are redundant. The related experts can pick out the useful features, but it may be a difficult and time-consuming task, especially when the data pattern is not clear. Missing relevant features or leaving irrelevant features are both harmful for real applications. It can make the mining algorithm useless. In addition, the presence of irrelevant or redundant features will increase the storage space, which not only slow down the mining process, but also impoverishes the quality of the discovered knowledge poor.

The challenge of data reduction is in fact dual as feature and instance selection are closely related. Depending on the subset of instances considered, the relevant features may change. Different subsets of features may yield to different subsets of relevant instances. To be useful, data reduction methods need to address both instance and feature selection. This strategy achieves a considerable reduction in the training data while maintaining, or even improving, the performance of the data-mining algorithm. The need consists in simultaneously selecting relevant features, and in cleaning the database by reducing the number of patterns. There are two difficulties: The first one is to find trade-offs between removing too many correct patterns and leaving some small overlap among classes. The second difficulty is that it suffers from a high degree of scalability problems, even for medium-sized datasets. Evaluation of all the solutions via greedy approaches is impracticable in many cases. Feature selection has been far more investigated than instance selection, and few algorithms have been presented that could cope with the particular problem of dual selection.

This chapter deals with the data reduction problem for instance and feature selection methods in the context of supervised classification. Selection for regression tasks (see [59] and [8]) has been largely investigated but is not addressed in this chapter. Many studies related to instance and feature selection separately are available. Simultaneous selection approaches are specific but often involve techniques or strategies developed for single selection. In this chapter, we review single selection algorithms as well as dual selection. Particular attention will be paid to solutions that can be scalable.

The remainder of the chapter is organized as follows: An overview of supervised sampling and feature selection methods is provided in Sects. 4.2 and 4.3. Section 4.4 is dedicated to dual selection. Approaches dealing with simultaneous feature and instance selection are presented as well as pseudocodes of popular and more recent methods.

4.2 Supervised Instance Selection Algorithms

Instance reduction is a task in which the input is the set of all data and the output is a subset of inputs. This means instance reduction consists in choosing a subset of the total available data to achieve the original purpose of a data mining application and it works as if the whole data is used. There are numerous instance selection methods for classification. The goal of an instance reduction method is to obtain a subset $S \subset T$ such that it does not contain any superfluous instances, i.e., noisy or redundant data and $acc(S) \approx acc(T)$, where $acc(X)$ is the classification accuracy obtained using X as a training set. It can use various strategies such as boosting, sampling, prototype selection, and active learning. It has been found that instance selection not only reduces the size of the dataset, but it also improves the dataset quality by selecting outlying, noisy, contradictory, or simply unhelpful instances. Instance selection has also been applied to both regression and time series prediction. For example, the decrease instance size for kNN regression (DISKR) algorithm [93] allows decreasing the size of the training set for k nearest neighbors (kNN) regression. It first removes the outlier instances that impact the regressor performance, and then sorts the remaining instances by the difference on output among instances and their nearest neighbors.

4.2.1 Overview of Supervised Instance Selection

Depending on the order in which instances are processed, instance selection methods can be classified into five categories. If they begin with an empty set and they add instances to the selected subset, analyzing the instances in the training set, they are called incremental. Decremental methods, on the contrary, start with the original training dataset and they remove those instances that are considered superfluous or unnecessary. Batch methods are those in which no instance is removed until all of them have been analyzed; instances are simply marked for removal if the algorithm determines that they are not needed, and at the end of the process only the unmarked instances are kept. Mixed algorithms start with a preselected set of instances. The process then decides either to add or to delete the instances. Finally, fixed methods are a sub-family of mixed ones, in which the number of additions and removals are the same. This approach allows them to maintain a fixed number of instances (more frequent in prototype generation). Considering the type of selection, three categories may be distinguished. The criterion is mainly correlated with the points that they remove: either border points, central points, or otherwise. Condensation techniques try to retain border points. Their underlying idea is that internal points do not affect classification, because the boundaries between classes are the keystone of the classification process. Edition methods may be considered the opposite of condensation techniques, as their aim is to remove those instances that are not well-classified by their nearest neighbors. The edition process achieves smoother boundaries as well as noise removal. The edition nearest neighbors (ENN) algorithm by Wilson [98] represents this algorithm family. Between these two approaches are hybrid algorithms, which try to maintain or even to increase the accuracy capability of the dataset by removing both internal and border points. Evolutionary approaches for instance selection have shown remarkable results in both reduction and accuracy. However, their main limitation is their computational complexity. Many techniques have proven to be successful. Finally, most of them are content to employ more or less sophisticated heuristics that simply search a prohibitively large solution space.

The process of instance selection (see [42] for a comparison of popular approaches and [77] for a recent review) was first used for instance based classifiers, such as k-nearest neighbors (k-NN) and is based on k-nearest techniques [12]. Using these approaches, faster and less costly classifications were obtained by maintaining only certain necessary instances in the classifier's dataset. Pioneering works were carried out by Hart [48], Ritter et al. [84], Gatesgates1972reduced, and Wilson [98]. That is, because instance based classifiers perform calculations for each instance of a dataset every time a new classification is to be made, a smaller dataset would require a smaller amount of memory storage and a fewer number of calculations. The goals of the first instance selection algorithms were therefore to select the minimum number of instances required to maintain the current classification accuracy of a dataset. Condensed nearest neighbor (CNN) rule by Hart [48] starts a new dataset from one instance per class randomly chosen from the training set. After that each instance from the training set that is wrongly classified using the new dataset is added to this set. This procedure is very fragile with respect to noise and the order of presentation. Reduced nearest neighbor (RNN) by Gates [38] follows the same concept while starting from the original training set and rejecting only those instances that do not decrease accuracy. Instance Based (IB) methods [2] are proposed; they are incremental methods, IB_1 relates to the 1-NN rule; IB_2 selects the instances misclassified by 1-NN (as CNN); IB_3 is an extension of IB_2 where a classification record is used in order to determine the instances to be retained. This corresponds to instances such that their deletion does not impact the classification accuracy. Iterative case filtering (ICF) that was proposed by Brighton and Mellish [13] is based on the concept of local set (see algorithm 5). Other methods related to k-NN are those proposed by Wilson and Martinez [99], namely the family of decremental reduction optimization procedure ($DROP$) methods from $DROP_1$ to $DROP_5$. These methods are based on the concept of associates combined with that of nearest neighbors. The relation of associate is the opposite of nearest neighbor: an instance p that has q as one of its nearest neighbors is referred to as an associate of q. The set of nearest neighbors of one instance is called neighborhood. Then, the associates of an instance p are those instances such that p is one of their k nearest neighbors. The set of associates for each instance is a list with all instances that have that particular instance in their neighborhood. $DROP_1$ discards an instance p from T if the associates of p in S are correctly classified without p; through this rule, $DROP_1$ discards noisy instances since the associates of a noisy instance can be correctly classified without it but in $DROP_1$, when the neighbors of a noisy instance are first removed, then the noisy instance will not be discarded. In order to solve this problem, $DROP_2$ is similar to $DROP_1$ but the associates of an instance are searched in the whole training set, that is, p is deleted only if its associates in T are classified correctly without p. $DROP_3$ and $DROP_4$ first discard noisy instances using a filter similar to ENN and then they apply $DROP_2$. $DROP_5$ is based on $DROP_2$ but it starts discarding the nearest enemies (nearest instances with a different class) in order to smooth the decision boundaries. Another way to find suboptimal solutions is the sequential search approaches. A decremental algorithm was proposed in [73]. Floating search methods [82] are more efficient but also more computationally expensive, the idea being the hybridization of forward and backward selections. They consist in applying after each forward/backward step a number of backward/forward steps as long as the quality subsets are better than the previously evaluated ones. The restricted floating object selection (RFOS) [74] method was proposed by the same authors to reduce the excessive computing time of these approaches by restricting the floating process. Local Density-Based Instance Selection ($LDIS$) [19] and eXtended Local Density-Based Instance Selection ($XLDIS$) [18] algorithms have been recently proposed. They evaluate the instances of each class separately and keep only the densest instances in a given neighborhood. These algorithms are simple and their performance in terms of the balance between accuracy and reduction is comparable to that of the other popular algorithms. Compared to $LDIS$, $XLDIS$ selects the instances that have the highest local density ordering in their k partial neighborhood (kPN). It avoids the inclusion of identical instances that have a high local density. $XLDIS$ keeps a list of instances that can be avoided. Then $XLDI$ does not need to check if they should be analyzed as candidates in the selection process, thereby reducing the lower computational time. In addition to the popular algorithms above, it should be mentioned that evolutionary algorithms which are a more usual process for dual selection (see Sect. 4.4.1) have also been used for instance selection as recently in [27]. They are rather efficient but less tractable than more classical methods.

4.2.1.1 Random Approaches and Genetic Algorithms

In addition to traditional approaches, it is worth mentioning that the majority of instance selection methods for training set selection can find an acceptable set of training instances utilizing random search with evolutionary algorithms[35, 60]. Evolutionary algorithms are a popular search technique because they are capable of taking into consideration the particular bias of a specific classification algorithm, and as such, lead to a good selection of training instances for that classifier. Most instance selection methods for training set selection are designed for neural networks and decision trees, but can be adapted for a variety of different classification algorithms that learn a classification scheme from a training dataset. More details are given in Sect. 4.4, on dual selection. Recently, a genetic-based prototyping algorithm [65] has been developed to obtain optimal prototype from images regarding the problem of image annotation. Then, for a given query image, its neighbor images are retrieved from the optimal prototype gained, and its candidate tags are generated using methods such as voting.

4.2.1.2 Instance Selection and Graph

The decision tree classifier is another popular choice for training set instance selection because unhelpful instances in the decision tree's training dataset cause the structure of the tree to grow unnecessarily [71, 89]. This overfitting problem defeats the purpose of the decision tree by hiding or confusing the discovered knowledge in a large and uninterpretable tree structure that has poor generalization abilities. Performing instance selection often results in a smaller and more interpretable tree, an indication that the unhelpful instances have not been selected. In [11, 15], the strategy consists in evolving the instances contained in a subset of the training dataset in the hopes of finding a collection of instances that adequately describe the full dataset. Endou and Zhao [30], as well as Cano et al. [17], evaluated the fitness of the selected training dataset through the construction and evaluation of a decision tree from the training dataset, and both of these methods are successful in reducing decision tree size while still maintaining acceptable, if not improved, levels of accuracy.

4.2.1.3 Performances via Reduction

Instance selection not only reduces the size of the dataset, but it also improves the dataset quality by selecting not to maintain outlying, noisy, contradictory, or simply unhelpful instances [77, 89, 106]. As a consequence of instance selection's ability to improve the quality of a dataset, it is now increasingly also being used to select good training datasets for learning a classification model, such as with the decision tree or neural network classifiers. This area of instance selection is different than early methods used for instance based classifiers because the goal is no longer data reduction but it is rather to maximize classifier accuracy. However, a good selection of instances for an instance based classifier may not lead to the best training dataset for another type of classifier, indicating that methods that incorporate the intended classifier in the learning process are warranted [75, 76]. Then, using the intended classifier in instance selection procedures can be beneficial because instances that obscure an advantageous structure or relationships with respect to the classifier's bias can be removed. This makes learning good boundaries between class values easier for the classifier. Therefore, with the majority of instance selection algorithms being developed for instance based classifiers, effort should be made to develop instance selection algorithms that are applicable to other classification methods used in practice.

4.2.2 Popular Algorithms

In this section, the pseudocodes of several popular algorithms are provided. Methods based on evolutionary algorithms are not included as these methods are conceptually close to the ones presented for dual selection in Sect. 4.4.2.

Wilson edition as a pioneer Edition nearest neighbors (ENN) algorithm is presented in algorithm 1. It follows a decremental process based on the k nearest rule. Let k be the number of nearest neighbors and $\{x_{(1)}, x_{(2)}, \ldots, x_{(n-1)}\}$ be the permutation of the elements of $X \setminus \{x\}$, such that:

$$||x_{(1)} - x|| \leq ||x_{(2)} - x|| \leq \ldots ||x_{(n-1)} - x|| \tag{4.1}$$

The k-nearest neighbors of the x item are the set defined as:

$$N(x) = \{x_{(1)}, x_{(2)}, \ldots, x_{(k)}\} \tag{4.2}$$

The training set $T = [X, C]$ includes $X_{n,p}$ and $C_{n,1}$ that are, respectively, the data matrix (n patterns in a p-dimensional space) and the corresponding class column vector label of dimension n. ENN acts on T as a noise filtering to provide the output $S = [X', C']$ ($X_{n',p}$ and $C_{n',1}$), where $n' \leq n$.

Algorithm 1 ENN [98]

Input: training set T (data and labels), k (number of nearest neighbors)
Output: S (selected subset $S \subset T$)
$S = T$
for all $(x \in T)$ **do**
    Find $N(x)$, the k nearest neighbors of x {Eq. (4.2)}
    Compute $\widehat{C(x)}$, the most frequent label in $N(x)$, ($C(x)$ *is the category of* x)
    if $C(x) \neq \widehat{C(x)}$ **then**
        $S = S \setminus \{x\}$ {if x is misclassified using its k nearest neighbors, it is removed from S.}
    end if
end for

Note that there are several extensions such as the Multi-Edit (the process of ENN is repeated as long as any changes are made in the selected set) and citation editing algorithms [29].

Instance based learning (IBL) algorithms classify instances based on the classification of their most similar neighbors. IB_3 [2] is an incremental algorithm that improves the preliminary versions 1 and 2. It deals with noise, keeps only good classifier data points, and discards instances that do not perform well. Instance x from the training set is added to a new set S if the nearest acceptable instance in S has a different class to x. An acceptable instance y is when $Ca(y)$ the ratio between the number of instances correctly classified and the total number of instances classified with y ($Ca(y) = correct[y]/classified[y]$) is more than a threshold θ_1. If there are no acceptable instances a random one is used. IB_3 employs a "wait and see" evidence gathering method to determine which saved instances are expected to perform well during classification. IB_3 maintains a classification record related to the past classification performance of each instance. Information for each instance y is the number of classifications done as well as the correct ones giving $Ca(y)$. IB_3 also uses a test based on a threshold θ_2 to determine which instances are believed to be noisy. If $(Ca(y) < \theta_2)$, y is removed from S.

Algorithm 2 IB$_3$ [2]

Inputs: T (training data with label), s (similarity function),θ_1 (threshold acceptable), θ_2 (threshold significantly bad)
Output: S (selected subset $S \subset T$)
$S = \varnothing$
for all $(x \in T)$ **do**
  correct[x]=0, classified[x]=0
  for all $(y \in S)$ **do**
    $Sim(y) = s(x, y)$ *(similaritybetweenyandx)*
  end for
  if $(S = \varnothing)$ **then**
    $S = S \cup \{x\}$
    GOTO: 4 {*at the first iteration* $S = \varnothing$}
  end if
  if $(\exists y \in S \mid Ca(y) \geq \theta_1$ **then**
    $y_{max} = \underset{y}{argmax}(Sim(y))$ {(*the most similar y from x among the acceptable y*)}
  else
    Select y_{max} randomly in S as the most similar instance to x
  end if
  if $(C(x) = C(y_{max}))$ **then**
    $Classification = 1$
  else
    $Classification = 0$
    $S = S \cup \{x\}$ (*x becomes a potential selected instance*)
  end if
  for all $(y \in S \setminus \{x\})$ **do**
    if $(Sim(y) \geqslant Sim(y_{max}))$ **then**
      $correct[y] = correct[y] + Classification$
      $classified[y] = classified[y] + 1$
      $Ca(y) = correct[y]/classified[y]$ {Update y's classification record}
      if $(Ca(y) < \theta_2)$ **then**
        $S = S \setminus \{y\}$ {*y is considered as noisy*}
      end if
    end if
  end for
end for

The $DROP_3$ algorithm, the most powerful one in the $DROP$ family method, is presented in algorithm 3. Let us define $A(x)$ as the associate of x:

$$A(x) = \{y \mid x \in N(y)\} \tag{4.3}$$

where $N(y)$ defines the nearest neighbors of y.

$A(x)$ is then the set of instances for which instance x is one of the k nearest neighbors. $DROP_3$ starts the process by running the ENN algorithm. In the first step, line 5, as $y \in N(x)$ x is added to $A(y)$. But the converse does not hold. The purpose of line 9 is to add y to $A(x)$ because $x \in A(y)$ but $x \notin N(y)$. This situation is illustrated in Fig. 4.1. Then the idea is to remove instance x from the training set if its presence does not improve or change the classification of instances from $A(x)$. The algorithm is incremental and modifies at each step the associate and neighbor sets of the remaining instances.

Algorithm 3 Drop3 [99]

1: Input: T (training set with label), k (number of nearest neighbors)
2: Output: S (selected subset $S \subset T$)
3: S = ENN(T,k) {Algorithm 1 Noise filtering}
4: **for all** $(x \in S)$ **do**
5:     $A(x) = \{y \mid x \in N(y)\}$ {Eq. (4.3), set of associates of x}
6: **end for**
7: **for all** $(x \in S)$ **do**
8:     **for all** $(y \in A(x))$ **do**
9:         $N(y) = N(y) \cup \underset{x \notin N(y)}{\{x\}}$ {Add x to each of its neighbors' list of associates}
10:     **end for**
11: **end for**
12: **for all** $(x \in S)$ **do**
13:     let $A_{with}(x) = \{y \in A(x) | C(y) = \widehat{C(x)}\}$ {list of associates of x classified correctly with x as a neighbor}
14:     $A_{without}(x) = \{y \in A(x) | C(y) = \widehat{C(x)}, N(y) = N(y) \setminus \{x\}\}$ {list of associates of x classified correctly without x}
15:     **if** $(|A_{without}(x)| \geq |A_{with}(x)|)$ **then**
16:         $S = S \setminus \{x\}$ {remove x from S}
17:         **for all** $(y \in A(x))$ **do**
18:             $N(y) = N(y) \setminus \{x\}$ {Remove x from y's list of nearest neighbors}
19:             find $u = \underset{z \in S \setminus N(y)}{argmin}(d(y, z))$ {Find a new nearest neighbor for y}
20:             $N(y) = N(y) \cup \{u\}$ {add u to the neighbor list of y}
21:             **for all** $(z \in N(y))$ **do**
22:                 $A(z) = A(z) \cup \{y\}$ {Add y to its new neighbor's list of associates}
23:             **end for**
24:         **end for**
25:     **end if**
26: **end for**
27: return S

The **Fast Condensed Nearest Neighbor** $(FCNN)$ rule algorithm (see algorithm 4) is based on the Voronoi cell concept. $FCNN$ selects points very close to the border decision. It is independent of the order, and has a quadratic complexity. S is a training set consistent subset of T for the NN rule if for each element x of S, $VorEn(x, S, T)$ is empty. $VorEn(x, S, T)$ is the set of Voronoi enemies of x in T with respect to S.

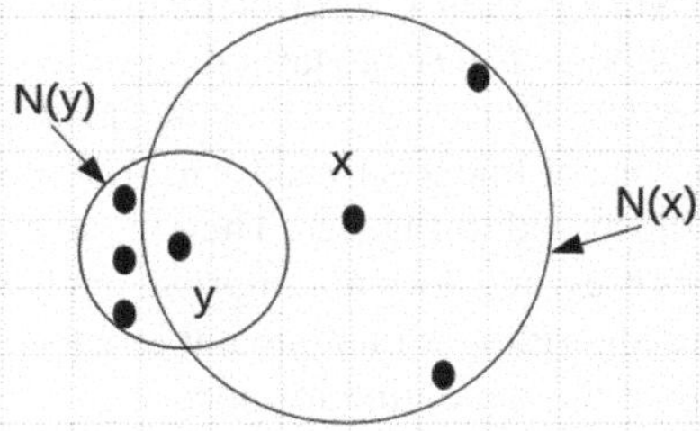

Fig. 4.1 Neighbors and associates

$$VorEn(x, S, T) = \{y \in Vor(x, S, T) \mid C(x) \neq C(y)\} \tag{4.4}$$

where $Vor(x, S, T)$ is the set of the elements of T that are closer to x than to any other element of S. $Vor(x, S, T)$ is a general notion that can be adapted through the k nearest neighbors rule. The algorithm initializes the subset S with a seed element from each class label of the training set T. The centroid of each class is the instance which is closest to the geometrical center of the class region. The algorithm is incremental. During each iteration of the algorithm, for each point x in S, the set of points y of T belonging to the Voronoi cell of x but having a different class label (i.e., $VorEn(x, S, T)$) is selected and added to the subset A. The algorithm stops when A is empty meaning that no further point can be added to S, that is, when T is correctly classified using S, i.e., all instances in T are correctly classified using S.

Algorithm 4 FCNN [6]

Input: T (training set with label)
Output: S (selected subset $S \subset T$)
$S = \varnothing$
$A = \varnothing$
for all (label l) **do**
  $L = \{x \mid C(x) = l\}$
  $y = \underset{x \in L}{argmin}(d(x, gc(L)))$ {$gc(L)$ is the geometric center of L}
  $A = A \cup \{y\}$
end for
while ($A \neq \varnothing$) **do**
  $S = S \cup A$
  $A = \varnothing$
  for all (x ⊂ S) **do**
    $A = A \cup VorEn(x, S, T)$ {Eq. (4.4). Add the enemies of x located in its Voronoi region}
  end for
end while

Iterative Case Filtering (ICF) was proposed by Brighton and Mellish [13]. The key concept is the local set (LS) for an instance x, $LS(x)$. It is the set of items in the same class of x located in the largest homogeneous hypersphere centered on x. Let $ne(x)$ be the nearest neighbor of x with a different label, i.e., the nearest enemy of x:

$$ne(x) = \underset{x' \mid C(x') \neq C(x)}{argmin} \; d(x, x') \tag{4.5}$$

The distance between x and $ne(x)$ is the radius of the hypersphere that defines the local set:

$$LS(x) = \{x' \in T \mid d(x, x') < d(x, ne(x))\} \tag{4.6}$$

The LS is used to define the *coverage* of an instance x: *coverage* is the set of elements that include x in their LS.

$$cover(x) = \{x' \in T \mid x \in LS(x')\} \tag{4.7}$$

The first step of the algorithm consists in removing noise from the neighborhood using the Wilson Edition algorithm (Algorithm 1). Then the algorithm iteratively removes items for which $|LS(x)| \geq |cover(x)|$. The idea is to keep only the instances that are useful to another one.

Algorithm 5 ICF [13]

Input: T (training set with label), k (number of nearest neighbors)
Output: S (selected subset $S \subset T$)
$S = ENN(T, k)$ {Algorithm 1. Noise filtering}
repeat
  for all $(x \in S)$ **do**
    $LS(x)$ {Eq. (4.6)}
    $cover(x)$ {Eq. (4.7)}
    Rem[x] = **false**
  end for
  progress = False
  for all $(x \in S)$ **do**
    if $(|LS(x)| \geqslant |cover(x)|)$ **then**
      Rem[x]=**true**
      $Progress$ = **true**
    end if
  end for
  for all $(x \in S)$ **do**
    if (Rem[x]=**true**) $S = S \setminus \{x\}$
  end for
until $Progress$ = **false**

The **Local Set-based Smoother** ($LSSm$) algorithm [62] is an Edition method based on the notion of usefulness ($u(x)$) and harmfulness ($h(x)$) of a given instance x. $u(x)$ is the number of instances having x among the members of their LSs and $h(x)$ is the number of instances having x as the nearest enemy:

$$u(x) = |LS(x)| \tag{4.8}$$

where $LS(x)$ is the Local Set of x.

$$h(x) = |\{y \mid ne(y) = x\}| \tag{4.9}$$

where $ne(x)$ is the enemy class of x. The idea behind the algorithm (see algorithm 6) is that a harmful instance misclassifies more instances than those that it correctly classifies. The key point is that the algorithm focuses on the influence of the instance in the neighborhood rather than the influence of the neighborhood for the instance. The algorithm consists in removing instances with a harmfulness greater than a usefulness. In contrast to most editing techniques, the methods avoid removing instances close to the border if the border is smooth. This is due to their closeness to instances of the other classes. On the contrary, it is likely to remove noisy instances or overlapped instances as they generally have a large harmfulness and a very low usefulness.

Local Set Border Selector ($LSBo$) is an instance selection algorithm derived from $LSSm$. The heuristic criterion of LSBo relies on the fact that instances which are near the class borders tend to have smaller LSCs than those located farther away. An instance is a border instance if it has a local set cardinality among the members of its local set. Then, the LSs in T' ($T' = LSSm(T)$) are computed, and the instances are sorted in ascending order by their LSC. Due to this sorting, the loop below starts processing the instances which are closer to the class boundaries. The loop verifies for each instance x in T if any member of its LS is contained in S, thus ensuring the proper classification of x. If that is not the case, x is included in S to ensure its correct classification.

Algorithm 6 LSSm [62]

```
Input: T (training set), k (number of nearest neighbors)
Output: S (selected subset S ⊂ T)
S = ∅
for all (x ∈ T) do
    Compute LS(x) {Eq. (4.6)}
end for
for all (x ∈ T) do
    Compute u(x) {Eq. (4.8)}
    Compute h(x) {Eq. (4.9)}
    if (u(x) ≥ h(x)) then
        S = S ∪ {x}
    end if
end for
```

Algorithm 7 LSBo [62]

```
Input: T (training set), k (number of nearest neighbors)
Output: S (selected subset S ⊂ T)
S = ∅
T' = LSSm(T) {Algorithm 6}
for all (x ∈ T') do
    Compute LS(x) {Eq. (4.6)}
end for
Sort T' according to |LS(x)| in increasing order
for all (x ∈ T') do
    I = LS(x) ∩ S
    S = S ∪ I
end for
```

$LDIS$ is an incremental algorithm that adopts the notion of local density $Dens(x, P)$:

$$Dens(x, P) = -\frac{1}{|P|}\sum_{y \in P} d(x, y) \tag{4.10}$$

where d is a given distance.

This definition was kept in the algorithm description to be in line with the original algorithm. The proposed formalism is however questionable as a density is not negative.

$LDIS$ also adopts the concept of partial neighborhood N^p. $N^p(x)$ is the set of the k nearest neighbors of x that has the same class of x. Let $T_C(x) = \{y \mid \underset{C(x)=C(y)}{y \in T}\}$. $N^p(x)$ is obtained in a similar way to $N(x)$ considering $T_C(x)$ as the reference set instead of T ($T_C(x) \subset T$).

$$N^p(x) = N_{T_C(x)}(x) \tag{4.11}$$

$N(x)$ being the neighborhood set of x regarding its k nearest neighbors.

$LDIS$ analyzes the instances of each class separately. It preserves only the densest instances in a given neighborhood. It verifies, for each x of a given class l, if there is some instance y $\in N^p(x)$ (Eq. (4.11)), such that its density is superior to that the one of x (related to class l). If this is not the case, this means that x is the locally densest instance in its partial k-neighborhood and, due to this, x is included in S.

Algorithm 8 LDIS [19]

Input: T (training set with label), k (number of nearest neighbors)
Output: S (selected subset $S \subset T$)
$S = \emptyset$
for all (label l) **do**
  for all $(x \mid C(x) = l)$ **do**
    $FoundDenser = False$
    for all (y $\in N^p(x)$) {Eq. (4.11) } **do**
      $T^l = \{z \in T \mid C(z) = l\}$
      if (Dens(x,T^l) $\leqslant$ Dens(y,T^l) **then**
        $FoundDenser = True$ {Eq. 4.10, Dens(x,T^l) local density of x limited to $y \in T^l$}
      end if
      if (foundDenser is False) **then**
        $S = S \cup \{x\}$
      end if
    end for
  end for
end for

4.2.3 *Large Scale Instance Selection*

The main drawback of instance selection methods is their complexity that is quadratic $o(n^2)$, where n is the number of instances or, at best, log-linear $o(n\ log\ n)$; thus, the majority of them are not applicable in datasets with hundreds or even many thousands of instances. Over the last few years, different approximations have been used to try to adapt instance selection methods to big/huge datasets.

4.2.3.1 State of the Art

One approach to deal with massive datasets is to divide the original problem into smaller subsets of instances, a process known as stratification. The underlying idea of these methods is to split the original dataset into disjoint subsets, then an instance selection algorithm is applied to each subset. Many state-of-the-art algorithms are due to Spanish teams [16, 25, 26]. One of the main interest of a stratification process is as follows. For an $O(n^2)$ instance selection algorithm, if the dataset is divided into n_s subsets of size s, $n_s = n/s$, the algorithm must be applied n_s times. The complexity of a stratification approach is highly reduced as $O(n_s(s^2)) \ll O(n^2)$. The very first stratification proposal [16] consisted in splitting the original dataset into disjoint strata (groups or sets of instances) with the same class distribution as the original one. The scaling-up benefits can be tuned by varying the size of each stratum. Moreover, the stratification process is suitable for boosting any other method. These preliminary works were improved in [34]. Divide and conquer algorithms are also commonly used in the instance selection area. The principle underlying these algorithms [92] can be simply stated: if the problem posed by a given input is sufficiently simple, it is solved directly; otherwise, it is decomposed into independent subproblems, the subproblems are solved, then the results are composed. The process of decomposing the input problem and solving the subproblems gives rise to the term "divide and conquer" although "decompose, solve, and compose" would be more accurate. This concept was applied in [25] to manage instance selection (see algorithm 10). The study in [53] proposed a method based on a recursive application of instance selection to smaller datasets. The dataset is subdivided recursively into smaller subsets to filter out the less useful internal points. The prototypes which result from each subset are then coalesced, and processed again by the instance selection algorithm to yield more refined prototypes. After the recursive subdivision, the smaller subsets are reduced with any traditional instance selection algorithm. The resulting sets of prototypes obtained are, in turn, gathered and processed at the higher level of the recursion to yield more refined prototypes. This sequence of divide-reduce-coalesce is invoked recursively to ultimately yield the desired reduced instances. In this manner, prototypes which are in the interior of the Voronoi spaces, and are thus ineffective in the classification, are eliminated at subsequent iterations of the instance selection algorithm. A direct consequence of eliminating the "redundant" samples in the computations is that the processing time is significantly reduced. One useful way to accelerate instance selection methods and to be able to cope with massive datasets is to adapt them to parallel environments [26]. To do so, the way that algorithms work has to be redesigned. The MapReduce paradigm offers a robust framework with which to process huge datasets over clusters of machines. Recently, two novel algorithms based on locality-sensitive hashing (LSH-IS-S and LSH-IS-F) presenting a o(n) complexity were proposed [8]. They make a particular use of hash functions. As a recall, hash functions present any functions that can be used to map data of arbitrary size to data of a fixed size. The idea with hashing is to turn a complex input value into a different value which can be used to rapidly extract or store data. LSH is used to find similarities between instances. The principle consists in making the instance selection on each of the buckets (results) that will be obtained by LSH when applied to all instances. LSH-IS-S completes the selection process in a single pass, analyzing each instance consecutively. It processes instances in one pass, so not all instances need to fit in memory. LSH-IS-F performs two passes: in the initial one, it counts the instances in each bucket; in the second, it completes the instance selection with this information. Both algorithms can be seen as incremental methods, due to the fact that the selected dataset is formed by successive additions to the empty set. However, the second one conforms more closely to batch processing because it analyzes the impact of the removal on the whole dataset.

4.2.3.2 Popular Algorithms

This section provides the pseudocode of popular instance selection algorithms that are scalable.

Democratic Instance Selection [34] is based on repeating several rounds r of a fast instance selection process. In each round, a partition splits the whole dataset into different disjoint subsets, also called bins. Different approaches are possible. The one-dimensional grand tour method [9] with several simplifications is suggested in the democratic algorithm. Each round on its own would not be able to achieve a good performance. However, the combination of several rounds using a voting scheme is able to match the performance of an instance selection algorithm applied to the whole dataset while considerably reducing the runtime of the algorithm. The algorithm updates an array of votes by increasing them by one if the instance has been selected. After performing a predefined number of rounds, the array of votes is used to determine, by means of a threshold, which instances should be either selected or removed. As the threshold of vote v cannot be preestablished (depending on the specific dataset) it is selected directly from the dataset in runtime by taking into account two criteria: training error, ε_t , and storage, or memory, requirements m. Both values must be minimized. A criterion $f(v)$ is defined, which is a combination of these two values:

$$f(v) = \alpha\varepsilon_t(v) + (1 - \alpha)m(v) \tag{4.12}$$

where m is measured as the percentage of instances retained, ε_t is the training error, and α is a value in the interval [0, 1] that measures the relative relevance of both values. After r rounds, each instance obtains a number of votes. The criterion $f(v)$ is processed for all the possible threshold values (in the range $[1, r]$) obtained in the different rounds. v is assigned to the value that minimizes the criterion.

The idea of the **Divide and Conquer** algorithm in [25] consists in splitting the whole dataset into disjoint sets. After the first batch of sets has been processed, the instances selected by the algorithm are joined, and the process starts again. This recursive instance selection proceeds until a certain reduction is achieved or any other stopping criterion is met. Fixing a number of iterations as stopping criterion for every dataset is possible but difficult. It depends on the specific features of each

Algorithm 9 Democratic instance selection [34]

Input: training set T, number of rounds r, threshold th, instance selection algorithm A
Output: S (selected subset $S \subset T$)
for all (rounds r) **do**
  Process $t_i \mid \cup t_i = T$ { Divide instance to n_s disjoint subsets t_i of size.}
  for all (subsets t_i of size n_s) **do**
    $s_i = A(t_i)$ {Apply A to t_i}
    Store votes of $t_i \setminus s_i$ {removed instance from t_i}
  end for
end for
$th = \underset{v\in[1,r]}{argmin}\ f(v)$ {Eq. (4.12) threshold of votes to remove an instance}
$R = \{x \in T \mid votes(x) \geq th\}$
$S = T \setminus R$ {Remove from S all the instances having votes $\geq th$}

problem. Thus, a cross-validation approach can be used. The training set is divided into two parts, using one of them for performing the instance selection algorithm and the other one for obtaining the validation error. The number of iterations is obtained as the last iteration before the validation error starts to grow.

Algorithm 10 Divide and conquer [25]

Input : the training set T, instance selection algorithm A, size s, stopping criterion
Output: the selected subset S ($S \subset T$)
$S = T$
Compute all $t_i \mid \bigcup_i t_i = T$ {Partition instances into disjoint subsets t_i of size s}
repeat
  for all ($t_i \in S$) **do**
    $s_i = A(t_i)$ {Apply A to t_i}
    $S = S \setminus t_i \cup s_i$ {Remove from S instances removed from t_i}
  end for
  Compute all $t_i \mid (\bigcup_i t_i) = S$ with $t_i = (\bigcup_j s_j)$ and $|t_i| \approx s$ {Fusion subsets s_j to obtain new subsets t_i of approximatively size s}
until stopping criterion

The **Federal Algorithm** in [26] presents a methodology for scaling up instance selection algorithms by means of a parallel procedure that performs instance selection on small subsets of the original dataset. The concept has some similarities with the democratic instance selection algorithm [34]. The idea consists in performing r partitions in disjoint subsets, so all instances are included exactly in r subsets. An instance selection algorithm (that is a parameter of the method) is applied to each subset concurrently. As for the democratic algorithm, the grand tour is suggested to generate the partition. The algorithm can be launched in parallel as most of the tasks can be performed concurrently reducing its complexity. As a key point for huge datasets, there is no need to have the whole dataset in memory. The instances that are selected to be removed receive a vote. After all the tasks are finished, the instances which have received a number of votes above a certain threshold are removed. As for the democratic selection algorithm, the election of the number of votes is based on two different criteria (Eq. (4.12)): training error ε_t, and storage, or memory, requirements m.

The criterion is evaluated in a federal way. These subsets are sent to the slaves which perform an evaluation of $f(v)$ for the subset using cross-validation and return the evaluation to the master. The master records the evaluations made by each slave and assigns the fitness, $f(v)$, averaged by the slaves to each possible value of v. Then, it assigns v to the value which minimizes the criterion. After that the instance selection is performed removing the instances whose number of votes is above or equal to the obtained threshold v.

Algorithm 11 Federal selection [26]

1: Input: Training set T, subset size s,number of rounds r, Instance algorithm A, number of processors p_r
2: Output: the selected subset S ($S \subset T$)
3: **for all** (rounds i) **do**
4: Compute $t_i[j]_{j \in [1,n_s]} \mid \bigcup_j t_i[j] = T$ {Divide instances into n_s disjoint subsets for each round i}
5: **end for**
6: Initialization: $i = 1$ and $j = 1$ {i refer to the round index, j refers to the subset index}
7: ** start the process by initiating a first task to each slave (processor)**
8: **for all** (processors p_r) **do**
9: *Master task*: send subset $t_i[j]$ to one Slave l
10: *Slave task*: $s_i[j] = A(t_i[j])$ {Apply A and send $A(t_i[j]) \setminus s_i[j]$ to Master}
11: **if** ($j = n_s$) **then**
12: $j = 1$ and $i = i + 1$
13: **end if**
14: $j = j + 1$
15: **end for**
16: **Continue the process of sending tasks to slaves until all t_i are managed**
17: **repeat**
18: *Master task*: Wait for a result for one slave w among p_r to finish
19: *Master task*: Store results from w { i.e. $t_{i_w}[j_w] \setminus s_{i_w}[j_w]$ the set of removed instances, i_w and j_w being the w indexes}
20: **if** (More subsets to process) **then**
21: *Master task*: Send Subset $t_i[j]$ to Slave w
22: *Slave Task* (w): process $s_i[j] = A(t_i[j])$, and send $A(t_i[j]) \setminus s_i[j]$ to the Master {the removed instances from w}
23: **if** ($j = n_s$) **then**
24: $j = 1$ and $i = i + 1$
25: **end if**
26: **end if**
27: $j = j + 1$
28: **until** (all $t_i[j]$ are performed: $i = r$ and $j = n_s${Master has received all the results from its slaves})
29: Apply the parallel loop (line 6 to 26) to perform ε and m {Each slave receives the removed instances and processes ε and m}
30: *Master task*: process $f(v)_{v \in [1,r]}$
31: $th = argmin_{v \in [1,r]} f(v)$ {Equation 4.12 threshold of votes to remove an instance}
32: $R = \{x \in T \mid votes(x) \geq th\}$
33: $S = T \setminus R$ {Remove from S all the instances having votes $\geq$ th}

4.3 Feature Selection Algorithms

The problem in feature selection (FS) can be easily stated as the search for a sufficiently reduced subset of, say, s features out of the total number of available ones, p, without significantly degrading (or even improving in some cases) the performance of the resulting classifier when using either set of features [44]. This search problem is driven by a certain measure of performance or criterion function which is used to assess the validity of each feature subset. This criterion has to be related to the final performance measure of the resulting classifier, i.e., its recognition rate. There are many potential benefits of variable and feature selection: facilitating data visualization and data understanding, reducing the measurement and storage requirements, reducing training and utilization times, defying the curse of dimensionality to improve prediction performance. Feature selection has been a fertile field of research since the 1970s in statistical pattern recognition, machine learning, and data mining, and widely applied to many fields such as text categorization, image retrieval, customer relationship management, intrusion detection, and genomic analysis. Feature selection is a process that selects a subset of original features. The optimality of a feature subset is measured by an evaluation criterion. As the dimensionality of a domain expands, the number of features p increases. The goal consists in constructing and selecting subsets of features that are useful to build a good predictor. This contrasts with the problem of finding or ranking all potentially relevant variables. Selecting the most relevant variables is usually suboptimal for building a predictor, particularly if the features are redundant. Conversely, a subset of useful features may exclude many redundant, but relevant, features. Finding an optimal feature subset is usually intractable and many problems related to feature selection have been shown to be NP-hard. For most problems it is computationally intractable to search the whole space of possible feature subsets. One usually has to settle for approximations of the optimal subset. Most of the research in this area is devoted to finding efficient search-heuristics.

4.3.1 Preliminaries: Basics of Information Theory

This section gives some elements of information theory on which many of the state-of-the-art algorithms related to feature selection are based. More details and theoretical concepts can be found in [14].

4.3.1.1 Entropy

The entropy of a discrete random variable X with probability mass function (pmf) $pX(x)$ measures the expected uncertainty in X.

$$H(X) = -\sum_{x} p(x)log(p(x)) \tag{4.13}$$

where $p(x)$ is the probability mass function. When X is discrete $p(x)$ is simply the ratio between the number of instants with x and the total number of instants. Dealing with feature selection problems, the features within these datasets have different characteristics, being binary, discrete or categorical, or continuous. The continuous features in feature selection approaches are generally discretized into δ intervals, using different methods [100].

In the case of continuous random variables, the summation is replaced by a definite integral and one talks about probability density function (pdf).

$H(X)$ is approximately equal to how much information is learnt on average from one instance of the random variable X. Entropy is always non-negative $H(X) \geq 0$ and $H(X) = 0$ iff X is deterministic.

With two random variables X, Y jointly distributed according to the pmf $p(x, y)$

$$H(X, Y) = -\sum_{x,y} p(x, y) \log(p(x, y)) \tag{4.14}$$

$H(X, Y)$ is named the joined entropy.

The conditional entropy of X given Y is

$$H(X|Y) = -\sum_{x,y} p(x, y) \log(p(x|y)) \tag{4.15}$$

4.3.1.2 Mutual Information

The concept of mutual information is intricately linked to that of entropy of a random variable. It quantifies the "amount of information" obtained about one random variable through observing the other random variable The mutual information I between two discrete random variables X, Y jointly distributed according to $p(x, y)$ is given by

$$I(X, Y) = -\sum_{x,y} p(x, y) \log \frac{p(x, y)}{p(x)p(y)} \tag{4.16}$$

$$I(X, Y) = H(X) - H(X|Y) = H(Y) - H(Y|X) = H(X) + H(Y) - H(X, Y) \tag{4.17}$$

$I(X, Y)$ is always positive.

4.3.1.3 Conditional Mutual Information

Let X, Y, Z be jointly distributed according to some p.m.f. $p(x, y, z)$. The conditional mutual information between X, Y given Z is

$$I(X, Y|Z) = -\sum_{x,y,z} p(x, y, z) \log \frac{p(x, y|z)}{p(x)p(y|z)} \tag{4.18}$$

4.3.1.4 Other Usual Notions

Let S_F be a set of selected features and f_i a feature candidate. Feature redundancy $R(f_i, S_F)$ is defined as the summation of the mutual information between a feature candidate f_i and each of the selected features.

$$R_f(f_i, S_F) = \beta \sum_{f_j \in S_F} I(f_i, f_j) \tag{4.19}$$

where β is a normalization factor ($|F|$).

Feature relevancy $R(f_i, C)$ expresses the relevance of f_i related to a class vector C.

$$R_f(f_i, C) = I(f_i, C) \tag{4.20}$$

This notion can be extended to the relevancy of a subset of m features related to C.

$$R_f(\{f_1 \dots f_m\}, C) = \sum_{i \in [1,m]} \sum_{\forall S \subseteq \{f_1 \dots f_m\}, |S|=i} I(S \cup C) \tag{4.21}$$

where $I(S \cup C) = \underset{(i<m)}{I(f_1, \dots f_i)}$

Interaction information is defined as follows:

$$I(f_i, Y, f_j) = I(f_i, Y) + I(Y, f_j) - I(Y, f_i, f_j) \tag{4.22}$$

where $I(Y, f_i, f_j)$ is the joint mutual information between Y and (f_i,f_j).

4.3.2 *Overview of Feature Selection*

There exists a vast amount of literature on feature selection [4, 63] as it is central in many areas involving classification problems [24]. It remains an open research area: feature selection is a difficult problem not only because of the large search space, but also because of feature interaction issues. Feature selection methods are often classified into two main categories (filter and wrapper) and two peripheral ones (hybrid and embedded). Filtering approaches [56] aim at selecting features independently of the learning algorithm. The wrapper approach uses a criterion that is dependent on the performances of the learning algorithm. The wrapper approaches considered in [55] are globally better but only at great computational expense. They aim at making the model performance estimation more reliable while improving the discrimination accuracy. Hybrid approaches attempt to take advantage of the two models by exploiting their different evaluation criteria in different search stages and using different techniques (see [23, 50, 52, 72]). Embedded approaches are more specific. Feature selection is intrinsic to the learning algorithm, which simultaneously learns the classifier and chooses a subset of features. Recently, other types of feature selection approaches such as ensemble feature selection and clustering-based feature selection methods have attracted researchers. In the big data context, data stream solutions are sometimes needed and hence online streaming feature selection. In [105], the authors present a method based on an adaptive density neighborhood relation, named OFS-density. Using the neighborhood rough set theory, OFS-density does not require the domain information before learning. A new adaptive neighborhood relation using the density information of the surrounding instances is presented. No parameters need to be specified in advance. The method can select features with a low redundancy using the fuzzy equal constraint.

4.3.2.1 Filtering and Wrapper Methods

A filter method selects subsets of variables as a preprocessing step, independently of the classifier used. It is based on the idea of relevance. Basically, the problem is to find the feature subset of minimum cardinality that preserves the information contained in the whole set of features with respect to the class variable C. This problem is usually solved by finding the relevant features and discarding redundant and irrelevant features. Filter methods are usually fast but not optimized as they are not related to a classifier. Many filter feature selection algorithms are based on information theory [14] and can handle high-dimensional datasets since they are fast and efficient. There are two main categories: ranking and subset techniques. The idea of ranking techniques is to order features according to importance scores that can be statistic, probabilistic, or even related to classification scores. Based on the use of different classifiers, it is shown in [70] that there is no best ranking index for different datasets and different classifiers accuracy. The function of the number of features used may significantly differ. The only way to be sure that the highest accuracy is obtained in practical problems is by testing a given classifier on a number of feature subsets, obtained from different ranking indices. With filter-based feature subset techniques, one evaluates the importance of each feature and selects a subset of relevant features. Measures can be correlation-based, distance-based, and consistency-based measures. For the second scores, the underlying techniques are rank search, best first search, and genetic search. Many supervised models have been proposed. Some algorithms as in [46, 54, 57] remain popular even though old. Correlation-based feature selection (CFS) is a subset selection method, and mainly uses heuristic approaches to evaluate the effect of a single feature corresponding to each category in order to obtain an optimal feature subset. ReliefF is extended from relief to support multiclass problems. Its main idea is to take Euclidean distance as the correlation index and then weight features according to how well they differentiate instances of different classes.

The idea behind wrapper methods is to select features according to the predictive power of a classifier when using the subsets. This means that the results are correlated to the classifier but are also dependent on it. The most problematic weakness is the requirement of a large amount of computation. One needs to define the way to search the space of all possible variable subsets and how to assess the prediction performance. A wide range of heuristic search strategies can be used. The most classical are forward and backward selections as well as hybrid strategies. In forward selection, one starts with an empty feature set and adds features at each step. In backward elimination, one starts with a full feature set and discards features at each step. Wrapper feature selection methods also include genetic algorithms (GA) [41], ant colony optimization (ACO) [1], particle swarm optimization (PSO) [5], and successive feature selection (SFS). The predictive power is usually measured on a validation set or by cross-validation. Note also in [40] a feature selection algorithm based on the FOA algorithm (forest optimization algorithm) [39] that has been adapted to discrete space.

In [64] a **unifying platform** is proposed for feature selection and generic algorithms for filter (algorithm 12), wrapper (algorithm 13), and hybrid methods (algorithm 14) are provided enabling a better understanding of dedicated algorithms. Filter algorithms are based on the following process. Given a training dataset T ($F = \{f_1, \ldots, f_p\}$) being the feature set), the algorithm starts the search from a given subset S_{F0} (an empty set, a full set, or any randomly selected subset) and searches through the feature space by a particular search strategy. Each generated subset S is evaluated by an independent measure M and compared with the previous best one. If it is found to be better, it is regarded as the current best subset. The search iterates until a predefined stopping criterion is reached. Typically, an independent criterion (without a mining algorithm) is used in algorithms of the filter model to evaluate the goodness of a feature: distance measures, information measures, dependency measures, and consistency measures. The stopping criterion can be related to the following items:

1. The search completes.
2. Some given bound is reached, where a bound can be a specified number (minimum number of features or maximum number of iterations).
3. Subsequent addition (or deletion) of any feature does not produce a better subset.

Wrapper algorithms are based on the same process except that the measure M is replaced by a mining algorithm A. The stopping criterion is generally related to the classification result: A sufficiently good subset is selected (e.g., a subset may be sufficiently good if its classification error rate is less than the allowable error rate for a given task).

4.3.2.2 Hybrid and Embedded Methods

Hybrid methods are based on strategies aiming at reaching the twofold objective of an efficiency comparable to that of wrapper methods and computation time comparable to that of filter ones [32]. They have received much attention as novel feature selection methods. In hybrid methods, a common way is to apply filter

Algorithm 12 Generic filter [64]

Inputs: the training set T ($F = \{f_1, ..., f_p\}$), initial subset S_{F_0}, independent measure M, δ a stopping criterion
Output: the selected feature subset S_F ($S_F \subset F$)
Initialize $S_{F_{best}} = S_{F_0}$
$\gamma_{best} = Eval(S_{F_0}, T, M)$ {Evaluate S_{F_0} by M.}
while (δ is not reached) {δ is related to M} **do**
    S_F=Generate(T,S_{F_best}) {generate a subset for evaluation}
    $\gamma = Eval(S_F, T, M)$ {evaluate the current subset by M.}
    if ($\gamma \geq \gamma_{best}$) **then**
        $\gamma_{best} = \gamma$
        $S_{F_best} = S_F$
    end if
end while
$S_F = S_{F_best}$

Algorithm 13 Generic wrapper [64]

Inputs: the training set T ($F = \{f_1, ..., f_p\}$), initial subset S_0, Mining algorithm A, δ a stopping criterion
Output: the selected feature subset S_F ($S_F \subset F$)
Initialize $S_{F_best} = S_{F_0}$
$\gamma_{best} = Eval(S_{F_0}, T, A)$ {Evaluate S_{F_0} by A}
while (δ is not reached) **do** {δ is related to A}
    S=Generate(T, S_{F_best}) {generate a subset for evaluation}
    $\gamma = Eval(S_F, T, A)$ {evaluate the current subset by A}
    if ($\gamma \geq \gamma_{best}$) **then**
        $\gamma_{best} = \gamma$
        $S_{F_best} = S_F$
    end if
end while
$S_F = S_{F_best}$

approaches as a preprocessing step and wrapper ones to finish the task. In [50], candidate features are first selected from the original feature set via computationally efficient filters. The candidate feature set is further refined by more accurate wrappers. In [7], the hybrid algorithms are based on the combination of a wrapper FS and rank-based filter methods. The wrapper FS is itself based on a binary differential evolution (BDE) algorithm [81].

Embedded methods are specific to a given classifier as the variable selection is implicitly integrated in the training process. These methods typically work by including in the objective function of the learning algorithm a sparsity-inducing regularizer or prior. These methods tend to be more computationally efficient than wrappers because they simultaneously integrate modeling with feature selection. This can be done, for instance, by optimizing a two-part objective function with a goodness-of-fit term and a penalty for a larger number of features. As with wrappers, the features selected by embedded methods are induction algorithm dependent. Embedded models embedding feature selection with classifier construction have the advantages of wrapper models as they include the interaction with the classification model and the advantages of filter models as they are far less computationally intensive than wrapper methods. Methods can be roughly divided into three different approaches: The first are pruning methods that first utilize all features to train a model and then attempt to eliminate some features by setting the corresponding coefficients to 0, while maintaining model performance. The second are models with a built-in mechanism for feature selection such as the C4.5 [86] algorithm. The third are regularization models with objective functions that minimize fitting errors and in the meantime force the coefficients to be small or to be exact zero. Features with coefficients that are close to 0 are then eliminated. Examples include Lasso, Elastic Net, and various decision tree based algorithms, e.g., CART, C4.5, and most recently, XGBoost [20, 94].

Algorithms 12 and13 are hybridized in a novel sequential forward hybrid algorithm (Algorithm 14) as follows: In each round for a best subset with cardinality c, it searches through all possible subsets of cardinality $c + 1$ by adding one feature from the remaining features. Each newly generated subset S with cardinality $c+1$ is evaluated by an independent measure M and compared with the previous best one. If S is better, it becomes the current best subset best at level $c + 1$. At the end of each iteration, a mining algorithm A is applied on the best subset at level $c + 1$ and the quality of the mined result is compared with that from the best subset at level c.

4.3.2.3 Feature Selection and Relevance

Feature selection is highly related to the notion of relevance. There are however several definitions [102] in the literature: relevance of one variable, relevance of a variable given other variables, relevance given a certain learning algorithm. Most definitions are problematic, because there are problems where all features would be declared to be irrelevant. Two degrees of relevance can be defined [10, 51]: weak and strong relevance. A feature is relevant if it is weakly or strongly relevant and irrelevant (redundant) otherwise. Relevance does not mean optimality of the feature set: classifiers induced from training data are likely to be suboptimal as there is no access to the real distribution of the data. The relevance does not imply that the feature is in the optimal feature subset. Even irrelevant features can improve a classifier's performance [45]. Relevance of one feature is relative (to a given algorithm) and not absolute. In addition, it does not take into account the interaction effects with other features [67]. Therefore, defining relevance in terms of a given classifier (and therefore a hypothesis space) would be better. Strongly relevant features provide unique information about C (the vector class), i.e., they cannot be replaced by other features. Weakly relevant features provide information about C, but they can be replaced by other features without losing information about C. Irrelevant features do not provide information about C, and they can be discarded without losing information. A good feature subset is one that contains features highly correlated with (predictive of) the class, yet uncorrelated with (not predictive of) each other.

Algorithm 14 Generic hybrid [64]

Inputs: the training set T ($F = \{f_1, ..., f_p\}$), initial subset S_{F0}, an independent measure M, a mining algorithm A
Output: the selected feature subset S_F ($S_F \subset F$)
Initialize $S_{F_best} = S_{F0}$
$c = c_0 = |S_{F0}|$ {feature number in $[1,p]$}
$\gamma_{best} = Eval(S_{F0}, T, M)$ {Evaluate S_{F0} by M}
$\theta_{best} = Eval(S_{F0}, T, A)$ {Evaluate S_{F0} by A}
for ($p_f = c + 1$ to p) **do**
  for ($i = 0$ to $p - c$) **do**
    $S_{F_best} = S_{F_best} \bigcup_{f_i \not\subseteq S_{F_{best}}} \{f_i\}$ {generate a subset of cardinality n_f for evaluation with feature f_i}
    $\gamma = Eval(S_{Fval}, T, M)$ {evaluate the current subset by M}
    $S'_{F_best} = S_{F_best}$
    if ($\gamma \geq \gamma_{best}$) **then**
      $\gamma_{best} = \gamma$
      $S'_{F best} = S_{F_{val}}$
    end if
    $\theta = Eval(S'_{F_{best}}, T, A)$ {evaluate the current subset by A}
    if ($\theta \geq \theta_{best}$) **then**
      $\theta_{best} = \theta$
      $S_{F_best} = S'_{F_{best}}, c = |S_{F_best}|$
    else
      return $S_F = S_{F_best}$ {no more classification improvement}
    end if
  end for
end for

4.3.2.4 Feature Selection Versus Feature Extraction

With the recent success of deep learning, we should mention the notion of feature extraction and highlight the difference with feature selection. Feature selection refers to selecting a portion of the original dimensions that are most important to the task. Feature extraction refers to extracting a new and smaller representation set from the original dimension space. Principal component analysis (PCA), linear discriminant analysis (LDA), or canonic correlation analysis (CCA) represents this category but also deep learning architectures too like convolution neural networks (CNN). Sometimes, feature extraction is difficult to explain as the link between features from the original feature space and new features is generally complex. It is then often not suitable for many practical situations. Feature selection is clearly superior in terms of better readability and interpretability than feature extraction. Convolution neural network (CNN) [88] has been very successful in image processing. It is a kind of feed-forward deep neural network with a convolutional structure that performs very well. CNN is composed of two parts: an automatic feature extractor and a trainable classifier. The model inputs images directly, global and local features are extracted via linear convolution layers. After convolution layers, the following usually is a non-linear polling layer. The polling layer can reduce the resolution of the extracted features, and there are two methods of polling layer that are usually used in the network. Average polling can be seen as a further feature extraction process to reduce computation. For the way of max polling, it can be seen as a feature selection to obtain the most important point of local features. The complete training process can be observed as a wonderful combination of feature extraction and feature selection. Today this task is internally done but the information is not factually accessible and interpretable. There is however hope with recent advances that CNNs or other deep learning architectures will be able to provide powerful instance and feature selection. The only remaining problem is the computational time of deep learning approaches. They are not appropriate for the timing objective.

4.3.2.5 Large-Scale Feature Selection

High dimensionality causes two major problems for feature selection. The crucial one is the so-called curse of dimensionality limiting the use of wrapper methods because of time complexity. In some application fields, another difficulty faced by feature selection with data of very large dimensionality can be the relative shortage of instances. Several authors proposed hybrid mechanisms to perform relevant selections while paying attention to the computation time [43, 101]. In [47] for example, a forward search approach is proposed that works in two steps to reduce the number of subset evaluations. In the first step, all attributes are ranked. This can be done either with a filter method or with the wrapper. In the second step the algorithm builds p attribute subsets: the first set is the top-ranked attribute, followed by the two top-ranked attributes, the three top-ranked attributes, and so on. These subsets are evaluated using the wrapper or a filter method that can evaluate sets of attributes. The authors used this technique to compare various filter techniques to the wrapper. With $2p$ evaluations, this algorithm known as rank-search is quite fast but chooses relatively large subsets. Rank-searched based instance (BIRS) [87] is one of the most popular approaches. An initial ranking is produced based on a filter method or the wrapper. The second step constructs attribute subsets and uses the wrapper for evaluation. The algorithm then starts with the top-ranked attribute and regards the remaining attributes in order of the ranking, but it only adds an attribute if it improves the current subset significantly. This method requires $2p$ evaluations, but generates smaller subsets than rank-search. A new hybrid feature selection algorithm has been recently proposed in [68]. Interaction information-guided incremental selection (IGIS) is computationally efficient with high accuracy rates for high-dimensional data. The proposed method employs interaction information to guide the search. Many individual features may be irrelevant for a class, but when combined together, they can interact and provide information that is useful for classification. IGIS sequentially adds one feature at a time into the currently selected subset, and adopts early stopping to prevent overfitting and speed up the search. The method selects only relevant and irredundant features that significantly improve the accuracy rates. In [70], state-of-the-art algorithms generally address generic problems. Some domains have however some specific problematics. It is the case for the big data bioinformatics area where the number of features can be very large. Main principles of feature selection and their recent applications in this field can be seen in [7, 96].

4.3.3 Popular Algorithms

This section provides several pseudocodes of popular feature selection algorithms.

Relief is an algorithm that takes a filter-method approach to feature selection. It was originally designed for binary classification problems with discrete or numerical features. Relief calculates a feature score for each feature which can then be applied to rank and select top scoring features for feature selection. Relief feature scoring is based on the identification of feature value differences between nearest neighbor instance pairs. At each iteration, one random instance x_i is selected in the training set T, and the feature vectors of the instance closest to x_i (by a given distance d) from each class. The closest same-class instance h is called "near hit," and the closest different-class instance m is called "near miss."

$$h(x_i) = \{y \in T \mid y = \underset{z \in T \setminus \{x_i\} \mid C(z)=C(x_i)}{argmin}\, d(x_i, z) \,\} \tag{4.23}$$

$$m(x_i) = \{y \in T \mid y = \underset{z \in T \setminus \{x_i\} \mid C(z)\neq C(x_i)}{argmin}\, d(x_i, z) \,\} \tag{4.24}$$

If for a given feature a difference is observed in a neighboring instance pair with the same class ("near hit") values, the feature score decreases. Alternatively, if a feature value difference is observed in a neighboring instance pair with different class ("near miss") values, the feature score increases. Relief is provided in Algorithm 15.

Algorithm 15 Relief [54]

1: Inputs: training data T ($F = \{f_1, ..., f_p\}$), r (number of random training instances), d (distance function)
2: Outputs: $W = \{W_1, ..., W_p\}$ scores that estimate the quality of the p features
3: $W_j = 0$ (for all features)
4: **for all** (r random trials) **do**
5: Randomly select a target instance x_i
6: **for all** (feature j) **do**
7: $W_j = W_j - \frac{d(x_i[j], h(x_i[j])}{r} + \frac{d(x_i[j], m(x_i[j]))}{r}$
{$h(x_i[j])$ and $m(x_i[j])$ are the nearest hit and miss: Eqs. (4.23) and (4.24)}
8: **end for**
9: **end for**

ReliefF evaluates the worth of a feature by repeatedly sampling an instance and considering the value of the given feature for the nearest instance of the same and different class. This attribute evaluation assigns a weight to each feature based on the ability of the feature to distinguish among the classes, and then selects those features whose weights exceed a user-defined threshold as relevant features. The weight computation is based on the probability of the nearest neighbors from two different classes having different values for a feature and the probability of two nearest neighbors of the same class having the same value of the feature. The higher the difference between these two probabilities, the more significant the feature is. Inherently, the measure is defined for a two-class problem, which can be extended to handle multiple classes, by splitting the problem into a series of two-class problems. It is described in Algorithm 16. Other extensions of relief have been proposed as in [22].

Algorithm 16 ReliefF [57, 58]

Input: training data T ($F = \{f_1, ..., f_p\}$), r (number of random training instances), k nearest neighbors, d (distance function)
Output: $W = \{W_1, ..., W_p\}$ scores that estimate the quality of the p features
$W_j = 0$ (for all features)
for all (r random trials) **do**
    Randomly select a 'target' instance x_i
    for all (feature j) **do**
        for all (neighbor k) **do**
            $$W_j = W_j - \frac{d(x_i[j], h^k(x_i[j])}{r} + \frac{d(x_i[j], m^k(x_i[j])}{r}$$
            $\{h^k(x_i[j])$ and $m^k(x_i[j])$ are the kth nearest hit and miss.$\}$
        end for
    end for
end for

Feature selection is usually based on mutual information criteria of max-dependency, max-relevance, and min-redundancy methods [78]. **Minimum redundancy and maximum relevance feature selection** (mRMR) is considered as one of the most powerful filters among the machine learning community. The general idea consists in minimizing the redundancy of features as well as maximizing the relevance regarding the class vector C. Many recent algorithms are based on mRMR. It is used to rank the importance of a set of features for a given classification task. This method can rank features based on their relevance to the target, and, at the same time, the redundancy of features is also penalized. The main objective is to find the maximum dependency between a set of features F and the class C, using mutual information I (Eq. (4.16)).

Algorithm 17 mRMR [78]

Input: training data T ($F = \{f_1, ..., f_p\}$, class variable C), number of wanted features n_f
Output: selected feature subset S_F ($S_F \subset F$)
for all (feature candidate f_i) **do**
    $R_l[i] = I(f_i, C)$
    {Equation (4.16): relevance of each feature}
    $R_d[i] = 0$
    for all (feature candidate f_j) **do**
        $R_d[i] = R_d[i] + I(f_i, f_j)$
        {redundancy of feature i with the other ones}
    end for
    $mRMR[i] = R_l[i] - R_d[i]$
    {a subtractive scheme is adopted}
end for
Select the n_f first according to mRMR scores.

Ranked-Searched Based Instance (BIRS) [87] is a popular feature selection algorithm. In the original algorithm features are ordered according to their individual accuracy rates, say the performance of a predefined classifier built with a single feature. In the second phase, the list of ranked features is processed only once from the best to the worst. Only the features that, when added to the currently selected feature set, improve the performance result will be kept. Any classification algorithm can be performed. The algorithm terminates when it reaches the end of the ranked list, and the currently selected feature set is returned. A modified version which was proposed in [69] is described in Algorithm 18. It was applied for predicting domain–peptide interactions. The difference is only in the first phase that uses the mRMR algorithm to rank all N features of the training set.

Algorithm 18 Modified BIRS [69]

Input: the training data T ($F = \{f_1, ..., f_p\}$)
Output: the selected feature subset S_F ($S_F \subset F$)
$Rank = mRMR(T)$
{Rank the features using the mRMR algorithm}
$BestScore = 0$
$BestSubset = \varnothing$
for all (ranked feature i) **do**
    $TempSubset = BestSubset \cup \{F_{Rank[i]}\}$
    $TempScore = WrapperClassif(TempSubset, C)$
    if ($TempScore \geqslant BestScore$) **then**
        $BestSubset = TempSubset$
        $BestScore = TempScore$
    end if
end for
$S = BestSubset$

Minimal Redundancy-Maximal New Classification Information (MR-MNCI) is a feature selection algorithm that integrates two groups of feature evaluation criteria. It is based on the observation that the methods that focus on minimizing feature redundancy do not consider new classification information and vice versa, thereby resulting in selected features with large amounts of new classification information but high redundancy, or features with low redundancy but little new classification information feature redundancy or maximizing new classification information. Concerning feature redundancy, the method adopts both class-dependent feature redundancy and class-independent feature redundancy. The proposed criterion consists of three terms, namely feature relevancy, new classification information, and class-independent feature redundancy.

Algorithm 19 MR-MNCI [33]

Input: the training data set T ($F = \{f_1, ..., f_p\}$, class variable C), threshold Th (number of wanted features)
Output: the selected feature subset S_F ($S_F \subset F$)
$F_T = F$
$u = 1$
$S_F = F_u = \underset{f_i \in F_T}{argmax}(I(f_i, C))$
{Init S_F with the feature j having the maximum $I(f_i, C)$}
for all (feature f_i) **do**
  Calculate $I(f_i, C)$
  {the mutual information between feature f_i and the class variable C}
end for
while ($u \leq Th$) **do**
  for all (candidate $f_i \in F_T$) **do**
    $RF_i = \sum_{f_j \in F_T} I(f_i, f_j)$
    {process feature redundancy f_i}
    $r_i = \frac{RF_i}{|F|}$ {normalize feature redundancy }
    $J(f_i) = I(f_i, Y) + \min_{f_j \in F_T} (I(f_i, C|f_j)) - r_i$ {Eq. (4.18) for $I((X, Y|Z)$}
  end for
  $F_u = \underset{f_i \in F_T}{argmax}(J(f_i))$ {Select the feature f_u with the best $J(f_i)$}
  $S_F = S_F \cup \{f_u\}$
  $F_T = F_T \setminus \{f_u\}$
  $u = u + 1$
end while

IGIS employs interaction information to guide the search. Many individual features may be irrelevant for the class, but when combined together, they can interact and provide information that is useful for classification. Based on interaction information, the proposed method conditionally adds one feature at a time into the currently selected subset and tests whether the resulting subset improves the performance significantly. Only relevant and irredundant features are thus selected. IGIS adopts early stopping to prevent overfitting and runs in linear time. It is composed of 4 steps.

Step1 IGIS starts with an empty set S and selects the first feature F_u from the full set F of p features that gives the largest mutual information between the feature and the class target C.

$$f_u = \underset{f_j \in F}{argmax}(I(f_j, C)) \tag{4.25}$$

The feature f_u is added to S and removed from S.

Step 2 A candidate feature is searched. The next candidate feature f_d is the one that maximizes the joint mutual information criterion

$$f_d = \underset{f_i \in F}{argmax}(I(f_i, C)) + \frac{1}{|S|} \underset{f_j \in S}{I(f_i, f_j, C)} \tag{4.26}$$

The first term measures the gain with C and the second the interaction between the currently selected feature and C.

Step 3 The candidate feature f_d is temporarily added to the current set S (i.e., $S \cup F_d$). The classification accuracy is computed. If it is improved significantly, go to step 4. Otherwise, f_d is not selected and then removed from the set F. If F is empty, terminate the algorithm. Otherwise, go to step 2.

Step 4 The classification accuracy is computed for the validation set with the subset $S \cup f_d$ using a given classifier. If the classification accuracy for the validation set does not decrease significantly (using a Student's paired left-tailed t-test at 0.1 level), permanently add f_d into the set S (i.e., $S = S \cup f_d$) and remove f_d from the set F, update the classification accuracy rates for the training and validation sets, and go to step 2. Otherwise, f_d is not selected and the algorithm terminates.

Algorithm 20 IGIS [68]

```
Inputs: the training set T (F = {f_1, ..., f_p}, class variable C), and a given classifier "Classifier"
Output: the selected feature subset S_F (S_F ⊂ F)
Divide T to create T_train and T_test.
F_T = F, f_u = argmax_{f_i ∈ F_T} (MI[i])
{select the first feature f_u, MI[i] = I(f_i, C)}
BestAccTrain = Classifier(T_train, T_train, f_u)
BestAccVal = Classifier(T_train, T_test, f_u)
S_F = {f_u}
F_T = F_T \ {f_u}
while (F_T ≠ ∅) do
    for all (f_i ∈ F_T) do
        Compute Interaction[i] {Eq. (4.26)}
    end for
    f_d = argmax_{f_j ∈ F_T} (Interaction[j])
    F_T = F_T \ {f_d}
    Stm = S ∪ {f_d}
    AccTrain = Classifier(T_train, T_train, Stm)
    if (AccTrain ≥ BestAccTrain) then
        AccVal = Classifier(T_train, T_test, Stm)
        if (AccVal ≤ BestAccVal) then
            break
        else
            BestAccTrain = AccTrain
            BestAccVal = AccVal
            S = Stmp
        end if
    end if
end while
```

4.4 Dual Selection

Dual selection can be performed separately via a sequential procedure. Another solution consists in managing the dual selection via a sole procedure. The authors in [95] tested different configurations. Based on their experiments, they concluded that performing feature selection first and instance selection second can make the classifiers provide slightly better classification results than performing instance selection first and feature selection second. However, the classifiers utilizing a combination of feature and instance perform slightly more poorly than the ones using feature selection or instance selection individually. On the other hand, in the large-scale experiments, the classifiers sometimes perform better based on a combination of feature and instance selection than those based on feature and instance selection alone. They therefore found it hard to figure out the winner of these four different data preprocessing steps since there is not a big difference between them. Consequently, the computational cost of training classifiers becomes another important indicator to assess these data preprocessing methods. The time complexity analysis shows that the combination of feature and instance selection greatly reduces the computational cost of training classifiers. As a result, it can be seen that the combination of feature and instance selection is a suitable solution for data preprocessing on large datasets.

4.4.1 Evolutionary Algorithms as a Usual Tool

Genetic algorithms (GAs) are one of the most widely used techniques for feature and instance selection [95], and can improve the performance of data-mining algorithms. They have the advantage of facing the so-called nesting effect of more classical methods in which a feature (instance) that is selected or removed cannot be removed or selected in later stages. In particular, [15] showed that better results can be obtained with GAs than with many traditional and non-evolutionary instance selection methods in terms of better instance selection rates and higher classification accuracy. Moreover, GAs have been shown to be suitable for large-scale feature selection problems. There have also been several studies in which GAs were used to perform both feature and instance selection tasks at the same time. Papers related to the simultaneous selection of features and instances are very few in number, and unquestionably the pioneers are those by Skalak [90, 91] and Kuncheva [61]. The idea was to define a chromosome representing the whole solution by encoding it in a string of bits, whose length is the sum of the number of available features and the number of patterns in the training set. In a chromosome a 1 for the ith feature or pattern stands for its selection, while a 0 means it is not taken into account. In [90] a single random mutation hill climbing procedure is proposed that gives predictive accuracy equal or superior to a basic nearest neighbor algorithm whose runtime storage costs were approximately 10–200 times greater. In [61], the authors conducted experiments where a GA was employed to simultaneously select suitable instances and features for a k-NN classifier. They used a fitness function that performs for the 1-*knn* rule and adds a penalty term as a soft constraint on the total cardinality of feature and instance sizes. They showed that a GA was an expedient solution compared to other traditional approaches. A GA method is proposed in [3] to simultaneously optimize feature weighting and instance selection for case-based reasoning in the bankruptcy prediction problem. Similarly, Ros et al. in [85] proposed a hybrid genetic approach, which treats feature and instance selection problems as a single optimization problem. The fitness function is different from the one proposed in [61] as penalty terms for feature and instance are distinguished. Moreover, several mechanisms are introduced in the GA procedure to improve the timing performance. In [31] the same authors propose an efficient nearest classifier that selects the most critical prototypes while discarding irrelevant and noisy features. GA and evolution strategies are combined in [83] to select instances and weight the features for the k-NN classifier. An intelligent genetic algorithm (IGA) was designed in [49] to tackle both instance and feature selection problems simultaneously by introducing a special orthogonal cross operator. The authors show that IGA performs better than the method developed in [61]. An evolutionary model based on cooperative coevolution is proposed in [28] to perform feature and instance selection in k-NN classification. This approach performs better than other evolutionary feature and instance selection methods over a wide range of datasets. It is worth mentioning that the drawback of GAs remains the difficulty of setting up and driving the algorithm to obtain good solutions in a reasonable time. This is especially critical when dealing with large databases as they are computationally expensive. They are therefore limited in the context of big data.

The **Simultaneous Selection Study** in [61] is a pioneering work dealing with simultaneous selection from which several more recent algorithms have been based. A selective (basic) genetic algorithm is proposed for simultaneous editing and feature selection. By itself, it is however not appropriate for large scale selection. The search space consists of 2^{n+p} elements. Each chromosome ch is represented by a binary string consisting of $(n + p)$ bits divided into 2 sets: the first n bits are used for the instance space, and the last p bits for the feature space. A population of P_s=10 chromosomes is taken in the original version. The fitness function is based on the k_{1NN} rule and a penalty term as a soft constraint on the total cardinality of n and p.

$$f(ch) = acc(ch) - \alpha\lambda(ch) \tag{4.27}$$

where $acc(ch)$ is the classification score (% of well-classified patterns), α the parameter that controls the balance between the criteria. $\lambda(ch) = \dfrac{n' + p'}{n + p}$, where (n', p') are the reduced cardinalities in chromosome x ($n' < n$ and $p' < p$).

Algorithm 21 Simultaneous selection [61]

1: Inputs: the training set T ($F = \{f_1, ..., f_p\}$, class variable C, population size p_s, Mutation probability P_m, Number of generations P_s.
2: Output: best chromosome (S_{XF}) {$\Leftrightarrow$ set S_{XF} ($S_{XF} \subset T$) reduced in the pattern and feature dimensions}
3: Initialize chromosomes (80% of bits are set to 1)
4: The whole population is taken as the "mating" set
5: **for all** (genetic generation) **do**
6: Select ($p_s/2$) couples randomly to produce p_s offsprings
7: Apply Crossover with probability 1 and Mutation with probability P_m.
8: Apply an Elitist selection:
(i) The p_s chromosomes and the current population are pooled.
(ii) The fittest p_s survive as the next population.
9: **end for**
10: Derive S_{XF} from the best chromosome

4.4.2 Large Scale Dual Selection and Popular Algorithms

Despite the extensive research efforts in the literature, most selection methods are restricted to batch learning settings. Existing solutions are still not feasible due to the high computation and memory cost in real-world applications. Existing approaches are unfortunately non-scalable when solving real-world applications with large-scale datasets that exceed the memory capacity. Another drawback is that batch learning methods usually assume that all training data and their full set of features have been made available prior to the learning process. There are two solutions: online selection and scalable approaches. Online feature selection has been studied for feature selection but not for dual selection. There are few studies dealing with simultaneous selection [21, 61, 85, 95] and very few dealing with scalability [36, 79, 103]. In [104], the authors addressed dual selection via in the case where the number of instances is fewer than the number of features. Different optimization techniques are used: Ant Lion Optimization (ALO), Grey Wolf Optimization (GWO), and a combination of the two (ALO-GWO) [66].

Pseudocodes of several dual selection algorithms are provided. The last one (Algorithm 24) provides a scalable strategy version. The others are based on GAs and share the same principle. They are not scalable by themselves but can be applied in a scalable way.

The principle of **Scalable Simultaneous Instance and Feature Selection** (SSIFS) [79] is similar to [61] but the evolution strategy and the chromosome are more sophisticated: The chromosome is enlarged to take into account weights associated with each feature and instance, following the principle of instance and feature weighting (see [97] for a review of weighting schemes). Given that there are n training instances with p features, the chromosome is of length $2n + 2p$. For each instance, the chromosome codifies a real value, its weight, and a binary value based on whether it is selected. It is the same for each feature: a unique weight is considered for the whole instance set. In synthesis, there are n+p weights and n+p bits. A differential evolution (DE) algorithm and cross generational elitist selection heterogeneous recombination and cataclysmic (CHC) genetic algorithm [15] are combined. Weights are evolved using a differential evolution algorithm, and selection (feature and instance) is evolved using a CHC genetic algorithm [15].

The fitness value, f, for an individual ch, is given thus by

$$f(ch) = \alpha_a acc(ch) + \alpha_i(1 - f_i(ch)) + \alpha_f(1 - f_f(ch)) \quad (4.28)$$

where $acc(ch)$ is the accuracy of the individual ch measured using a nearest neighbor rule, $f_i(ch)$ is the fraction of selected instances, $f_f(ch)$ is the fraction of selected features, and $\alpha_a + \alpha_i + \alpha_f = 1$.

The initial population is randomly created using a probability of 0.5 to select a feature or instance and a uniform distribution of weights in the interval [0, 1]. If any of the four tasks is not carried out, the corresponding weights are set to 1. Then, a new population is generated by differential mutation, recombination (via CHC), and selection. This cycle is repeated through a number of generations that is given as a parameter. Finally the individual with the best fitness in the last population is returned as the result of the optimization process.

Weighted Distances and Differential Evolution For instance weighting, the distance between an instance x and a query instance q is as follows:

$d(q, x) = w_x \sqrt{\sum_{j \in [1,p]} (q_j - x_j)^2}$, where w_x is the weight attached to x. For feature weighting, the weights are related to each feature that gives np weights. To simplify, if a unique weight for each feature is considered for the whole instance set, the weighting distance is as follows: $d(q, x) = \sqrt{\sum_{j \in [1,p]} w_j^2 (q_j - x_j)^2}$. Using weights proved to be efficient in accuracy and processing time in problems related to feature and instance selection. The different weights present novel parameters to be optimized. By combining the two weighting schemes a weighted distance depending both on instances and features can be obtained. Differential evolution operates through similar computational steps as those employed by a standard evolutionary algorithm (EA). However, unlike traditional EAs, the DE-variants perturb the current generation population members with the scaled differences of randomly selected and distinct population members. Therefore, no separate probability distribution needs to be used to generate the offspring.

Differential evolution generates new parameter vectors by adding the weighted difference between two population vectors to a third vector. The initial vector population is chosen randomly and should cover the entire parameter space. The basic strategy uses three different operators: mutation, crossover, and selection. Consider that an auxiliary population is generated from the current one. For mutation, a mutant chromosome for each target chromosome is generated, thus:

$$\underset{i \in [1, P_s]}{ch_i^m} = ch_{r1} + \gamma (ch_{r2} - ch_{r3}) \tag{4.29}$$

with random different indexes r_1, r_2, and r_3 and $\gamma > 0$. Crossover is used to increase the diversity of the perturbed parameter vectors. For each chromosome i, the crossover operation is applied to each of the component weights j ($j \in [1, n + p]$). It provides $ch_i^c = \{ch_{i1}^c, \ldots, ch_{i(n+p)}^c\}$ that is formed, thus:

$$\begin{cases} ch_{ij}^c = ch_{ij}^m \; if \; (r < CR \; or \; j = rnbr(i) \; else \\ ch_{ij}^c = ch_{ij} \end{cases} \tag{4.30}$$

where r is a random number in [0, 1] and $rnbr(i)$ is a randomly generated index related to chromosome i. It is in the range $[1, n + p]$ and ensures that ch_i^c gets at least one weight from the mutant chromosome ch_i^m. CR is a crossover constant, $CR \in [0, 1]$.

CHC Process In CHC, the chromosome of each individual has as many bits as instances plus features. A bit with a value of 1 means that the corresponding instance or feature is selected and a value of 0 means that the corresponding instance is not selected. Crossover is done via a half uniform crossover (HUX) [80]. This operator generates two offspring from two parents. Each offspring inherits the matching bits of the two parents ($of_{1i} = of_{2i} = b$ if $p_{1i} = p_{2i} = b$) and half of the non-matching bits ($p_{1i} \neq p_{2i}$) from each parent alternately ($of_{1i} = p_{2i}$ and $of_{2i} = p_{1i}$ half times). In SSIFS, this operation is applied to the $(n + p)$ bits in parallel to the differential evolution applied to the weights. $(P_s/2)$ pairs of parents are generated from the current population. If the Hamming distance between two parents is more than a threshold th, two children are generated using the HUX operator. If no crossovers are operated th is decreased by 1 until it becomes negative. If negative the population is regenerated and th is set to its initial value. This process is detailed in Algorithm 23. To simplify the reading of SSIFS, one defines the CHC process as follows: $G' = CHC_{HUX}(G)$ that produces a population G' from G the operation being restricted to only the binary part of the chromosomes.

The double crossover operation gives a population of trial vectors ch_i^m ($i \in [1, P_s]$). To decide whether it should become a member of generation $G + 1$, ch_i^m is compared to the target vector ch_i. If vector ch_i^m yields a better fitness function value than ch_i, then ch_i is set to ch_i^m; otherwise, the old value ch_i is retained.

Algorithm 22 SSIFS [79]

Inputs: training set T ($F = \{f_1, ..., f_p\}$, class variable C), population size P_s, γ, α_a, α_i, α_f, CR, stopping criterion.
Outputs: best chromosome $\{\Leftrightarrow$ set S_{XF} ($S_{XF} \subset T$) reduced in the pattern and feature dimensions$\}$
Set the generation number G = 0
Initialize a population of P_s chromosomes
for all (Chrom ch_i) **do**
    $\underset{u \in [1,p]}{f_u} = 1$ with $prob = 0.5$, $\underset{u \in [1,p]}{Wf_u} = random(U(0, 1))$
    {U is the Uniform distribution}
    $\underset{v \in [1,n]}{x_v} = 1$ with $prob = 0.5$, $\underset{v \in [1,n]}{Wx_v} = random(U(0, 1))$
end for
repeat
    Set $G = \{ch_1, ...ch_{Ps}\}$
    $G' = CHC_{HUX}(G)$ $\{G' = \{ch_1^c, ...ch_{Ps}^c\}$, only the binary part is managed$\}$
    for all (Chrom ch_i) **do**
        Mutation Step for the weights {Eq. (4.29): process the mutant vector ch_i^m with ch_{r1}, ch_{r2} and ch_{r3} and γ}
        Compute ch_i^c {Eq. (4.30): compute crossover for the weight part}
        if $f(ch_i^c) > f(ch_i)$ {ch_i^c comes from the CHC process and Differential Evolution} **then**
            $ch_i = ch_i^c$
        end if
    end for
    $G = G + 1$
until (Stopping criterion is satisfied)
Best Chrom $w = \underset{i \in [1,n]}{argmax}$ f(ch_i) {Eq. 4.28}
Derive S_{XF} from the best chromosome

Algorithm 23 Memetic algorithm [37]

Inputs: the training set T ($F = \{f_1, ..., f_p\}$, class variable C, population G size p_s, Crossover probabilities P_{FSLS}, P_{ISLS}, P_{IFSLS}, threshold function d_{min} and number of generation (stopping criterion).
Output: best chromosome (S_{XF}) {$\Leftrightarrow$ set S_{XF} ($S_{XF} \subset T$) reduced in the pattern and feature dimensions}
Initialize P_s chromosomes (ch) randomly (ch_i, $i \in [1, p_s]$)
$\underset{i\in[1,p_s)}{fitness[i]} = f(ch_i)$ {Evaluate fitness for all the chromosomes, Eq. (4.28)}
$th = d_{min}(n + p)$ {threshold function for the Hamming distance}
repeat
  Process $(P_s/2)$ pairs of parents (p_{1i}, p_{2i}) for crossover $\{p_{1i} \in G, p_{2i} \in G$ and $p_{1i} \neq p_{2i}\}$
  $G' = G = \{ch_1 ... ch_{P_s}\}$ {init G' with the current population}
  cross=0
  for all (pairs $p_{1i}, p_{2i} \in G$) **do**
    if $(d_{Hamming}(p_{1i}, p_{2i})/2 > th)$ **then**
      $(ch_{1i}, ch_{2i}) = HUX(p_{1i}, p_{2i})$, cross=cross+1
      {Apply HUX crossover to obtain children}
      $G' = \{ch_{1i}\} \cup G'$, $G' = \{ch_{2i}\} \cup G'$
    end if
  end for
  Process $\underset{i\in[1,|G'|)}{fitness[i]}$ {Evaluate fitness for all the $|G'|$ chromosomes}
  Process $\underset{i\in[1,|G'|]}{ch_i^1}$ = FSLS (ch_i,P_{FSLS}) $\{f(ch_i^1 \geqslant f(ch_i)\}$
  Process $\underset{i\in[1,[1,|G'|]]}{ch_i^2}$ = ISLS (ch_i^1,P_{ISLS}) $\{f(ch_i^2 \geqslant f(ch_i^1)\}$
  Process $\underset{i\in[1,[1,|G'|]}{ch_i^3}$ = IFSLS (ch_i^2,P_{IFSLS}) $\{f(ch_i^3 \geqslant f(ch_i^2)\}$
  $G = \underset{ch^3\in G'}{Selection}(ch^3, P_s)$
  {The fittest p_s among $[1, |G'|]$ are selected and survive as the next population $G = \{ch_1 ... ch_{P_s}\}$}
  if (cross=0) **then**
    th=th-1
  end if
  if $(th < 0)$ **then**
    G=Regenerate(G), $th = d_{min}(n + p)$
    {a restart process is operated}
  end if
until Stopping criterion is satisfied
$ch_{best} = \underset{i\in[1,p_s]}{argmax}\ fitness[i]$ {ch_{best} the best chromosome}
Derive S_{XF} from the best chromosome

The authors in [37] present a **memetic algorithm** based on a binary chromosome as in Algorithm 21. It is based on a fitness function similar to the one presented in Eq. (4.28). The algorithm has the following characteristics: (1) It is selective so as to obtain the next generation for a population of size P_s, the parents and the offspring are considered together, and the P_s best individuals are selected. (2) Premature convergence is prevented: only different individuals, separated by a threshold Hamming distance, d_{min}, are allowed to mate. The threshold distance is set to d_{min} = l/4, where $l = p + n$ is the length of the individual. If no individuals with a Hamming distance above d_{min} are found in a generation, and hence no matings are performed, then the threshold is decreased by 1. (3) The crossover operation is done via HUX. (4) Three local procedures are involved in the process: feature selection local search (FSLS) applying a backward selection, each feature being randomly tested; instance selection local search (ISLS) on the basis of an instance selection algorithm such as IB_3; and instance and feature selection local search (IFSLS) applying a sequential backward selection. $IFSLS$ begins with instance selection: each instance is removed and the fitness of the individual is reevaluated. If the fitness is equal to or better than the fitness with the instance selected, then the instance is removed permanently; otherwise, the instance is kept. Once the search in the space of instances is finished, the same procedure is repeated for the features.

Scalable Simultaneous Instance and Feature Selection Method (SSIFSM) [36] is a wrapper approach based on the divide-and-conquer principle combined with bookkeeping. It is one of the few papers that have addressed the simultaneous selection of features and instances allowing large scale dual selection. The training dataset T is divided into t disjoint subsets t_j (instance sampling) that are themselves divided in s subsubsets t_{ji} (feature sampling), with m features. An evolutionary algorithm for simultaneously selecting instances and features is applied on each subset t_{ji}. The fitness function of an individual ch depends on accuracy $acc(ch)$ and reduction $red(ch)$. It is driven by a parameter α as follows:

$$f(ch) = \alpha acc(ch) + (1 - \alpha) red(ch) \tag{4.31}$$

with $acc(ch)$ the classification score and

$$red(ch) = 1 - \lambda(ch) \tag{4.32}$$

The process is repeated r times, r being the number of rounds. At each round, the selections performed are recorded. Finally, the most relevant instances and features are retained via a voting process. The number of votes received by an instance is in the interval $[0; rs]$, as an instance is in s subsets in every round. Each feature is in t subsets each round. Thus, the number of votes is in the interval $[0; rt]$. The thresholds θ_i and θ_f are determined automatically using an evaluation function J as follows:

$$J(T(\theta_i, \theta_f)) = \beta acc(T(\theta_i, \theta_f)) + (1 - \beta) red(T(\theta_i, \theta_f)) \tag{4.33}$$

where $red(T(\theta_i, \theta_f))$ is the reduction achieved using threshold θ_i and θ_f to select $T(\theta_i, \theta_f)$, $acc(T(\theta_i, \theta_f))$ is the accuracy achieved with this selection using a k_{1nn} classifier and β to adjust the balance between the two criteria. β should be close to 1 ($\beta = 0.75$) in order to avoid a large reduction at the expense of poor accuracy. For every pair of thresholds (θ_i, θ_f), J is then evaluated in each instance partition and the pair of thresholds with the highest J is selected. The divide-and-conquer principle speeds up the execution of the algorithm.

Algorithm 24 SSIFSM [36]

1: Inputs: training set T ($F = \{f_1, \ldots, f_p\}$, class variable C), number of feature subsets s and number of rounds r
2: Output: S_{XF} $\{(S_{XF} \subset T)$ reduced in the pattern and feature dimensions$\}$
3: **for** ($i = 1$ to r) **do**
4: Process $t_i[j] \mid \cup t_i[j] = T$ {Divide instances into t disjoint subsets of approximately the same size}
5: **for** ($j = 1$ to t) **do**
6: $\cup t_i[j][u] = t_i[j]$ {Divide $t_i[j]$ into s disjoint subsets $t_i[j][u]$ with approximately the same number of features m}
7: **for** (u=1 to s) **do**
8: Apply instance selection algorithm to $t_i[j][u]$
9: Store votes of selected instances from $t_i[j][u]$
10: **end for**
11: **end for**
12: **end for**
13: Obtain thresholds of votes to keep an instance,θ_i , and a feature θ_f
14: $S_{XF} = \{x_i \in T \mid votes(x_i) \geqslant \theta_i$ and which features $j \mid vote(j) \geqslant \theta_i\}$

4.5 Conclusion

The rapid development of machine learning has led to significant advances in societal issues such as face recognition, disease diagnosis, speech recognition, image classification, and many other real-life problems. Sample selection and dimensionality reduction techniques are extremely important in large-scale data analysis, especially in machine learning.

The central point of feature and instance selection is approximation with the hope of achieving as good mining results as possible (within efficiency and timing criteria). The process consists in approximating the complete dataset with the selected instances. There are many ways of achieving approximation and it would be nice if there were a single general purpose selection method that guarantees a good performance in any situation. This chapter has reviewed the main state-of-the-art techniques of instance reduction and feature selection as well as some preliminary solutions of dual selection in the context of big data. It is clear that these two techniques have been applied widely, but as yet there is no universal method for sample selection and dimensionality reduction that can be applied to all problems. Sample and feature selection methods either use only one of these techniques or use all sequentially or simultaneously. Each specific problem typically adopts one unique method in order to improve on previous work. For very large or huge databases, the tendency is to apply stratification strategies combined with relevant algorithms that proved to be very efficient for small databases. To balance the trade-off between accuracy and reduction, evolutionary algorithms are often claimed as the most relevant even if they can have convergence issues. Despite good progress in solving feature and instance selection problems, more study is also welcomed to further optimize the solutions. In all the proposed methods, one should choose either computational feasibility or optimality. Further research is needed to develop more promising selection methods.

References

1. Aghdam, M.H., Ghasem-Aghaee, N., Basiri, M.E.: Text feature selection using ant colony optimization. Expert Syst. Appl. **36**(3), 6843–6853 (2009)
2. Aha, D.W.: Incremental constructive induction: an instance-based approach. In: Machine Learning Proceedings 1991, pp. 117–121. Elsevier, Amsterdam (1991)
3. Ahn, H., Kim, K.J.: Bankruptcy prediction modeling with hybrid case-based reasoning and genetic algorithms approach. Appl. Soft Comput. **9**(2), 599–607 (2009)
4. Al-Ani, A., Alsukker, A., Khushaba, R.N.: Feature subset selection using differential evolution and a wheel based search strategy. Swarm Evol. Comput. **9**, 15–26 (2013)
5. Amoozegar, M., Minaei-Bidgoli, B.: Optimizing multi-objective PSO based feature selection method using a feature elitism mechanism. Expert Syst. Appl. **113**, 499–514 (2018)
6. Angiulli, F.: Fast condensed nearest neighbor rule. In: Proceedings of the 22nd International Conference on Machine Learning, pp. 25–32. ACM, New York (2005)
7. Apolloni, J., Leguizamón, G., Alba, E.: Two hybrid wrapper-filter feature selection algorithms applied to high-dimensional microarray experiments. Appl. Soft Comput. **38**, 922–932 (2016)
8. Arnaiz-González, Á., Díez-Pastor, J.F., Rodríguez, J.J., García-Osorio, C.: Instance selection of linear complexity for big data. Knowl.-Based Syst. **107**, 83–95 (2016)
9. Asimov, D.: The grand tour: a tool for viewing multidimensional data. SIAM J. Sci. Stat. Comput. **6**(1), 128–143 (1985)
10. Bell, D.A., Wang, H.: A formalism for relevance and its application in feature subset selection. Mach. Learn. **41**(2), 175–195 (2000)
11. Bennette, W.D.: Instance selection for simplified decision trees through the generation and selection of instance candidate subsets. Graduate Theses and Dissertations. 12084. https://lib.dr.iastate.edu/etd/12084, doi:10.31274/etd-180810-1522 (2011)
12. Bhatia, N. et al.: Survey of nearest neighbor techniques. arXiv preprint arXiv:1007.0085 (2010)
13. Brighton, H., Mellish, C.: Advances in instance selection for instance-based learning algorithms. Data Min. Knowl. Discov. **6**(2), 153–172 (2002)
14. Brown, G., Pocock, A., Zhao, M.J., Luján, M.: Conditional likelihood maximisation: a unifying framework for information theoretic feature selection. J. Mach. Learn. Res. **13**, 27–66 (2012)
15. Cano, J.R., Herrera, F., Lozano, M.: Using evolutionary algorithms as instance selection for data reduction in KDD: an experimental study. IEEE Trans. Evol. Comput. **7**(6), 561–575 (2003)
16. Cano, J.R., Herrera, F., Lozano, M.: Stratification for scaling up evolutionary prototype selection. Pattern Recognit. Lett. **26**(7), 953–963 (2005)
17. Cano, J.R., Herrera, F., Lozano, M.: On the combination of evolutionary algorithms and stratified strategies for training set selection in data mining. Appl. Soft Comput. **6**(3), 323–332 (2006)

18. Carbonera, J.L.: An efficient approach for instance selection. In: International Conference on Big Data Analytics and Knowledge Discovery, pp. 228–243. Springer, Berlin (2017)
19. Carbonera, J.L., Abel, M.: A density-based approach for instance selection. In: 2015 IEEE 27th International Conference on Tools with Artificial Intelligence (ICTAI), pp. 768–774. IEEE, Piscataway (2015)
20. Chen, T., Guestrin, C.: XGBoost: a scalable tree boosting system. In: Proceedings of the 22nd ACM SIGKDD International Conference on Knowledge Discovery and Data Mining, pp. 785–794. ACM, New York (2016)
21. Chen, Z.Y., Lin, W.C., Ke, S.W., Tsai, C.F.: Evolutionary feature and instance selection for traffic sign recognition. Comput. Ind. **74**, 201–211 (2015)
22. Chikhi, S., Benhammada, S.: ReliefMSS: a variation on a feature ranking ReliefF algorithm. Int. J. Bus. Intell. Data Min. **4**(3–4), 375–390 (2009)
23. Das, S.: Filters, wrappers and a boosting-based hybrid for feature selection. In: Proceedings of the Eighteenth International Conference on Machine Learning (ICML), vol. 1, pp. 74–81 (2001)
24. Dash, M., Liu, H.: Feature selection for classification. Intell. Data Anal. **1**(3), 131–156 (1997)
25. de Haro-García, A., García-Pedrajas, N.: A divide-and-conquer recursive approach for scaling up instance selection algorithms. Data Min. Knowl. Discov. **18**(3), 392–418 (2009)
26. de Haro-García, A., García-Pedrajas, N., del Castillo, J.A.R.: Large scale instance selection by means of federal instance selection. Data Knowl. Eng. **75**, 58–77 (2012)
27. de Haro-García, A., Pérez-Rodríguez, J., García-Pedrajas, N.: Combining three strategies for evolutionary instance selection for instance-based learning. Swarm Evol. Comput. **42**, 160–172 (2018)
28. Derrac, J., Triguero, I., García, S., Herrera, F.: Integrating instance selection, instance weighting, and feature weighting for nearest neighbor classifiers by coevolutionary algorithms. IEEE Trans. Syst. Man Cybern. B **42**(5), 1383–1397 (2012)
29. Devijver, P.A., Kittler, J.: Pattern Recognition: A Statistical Approach. Prentice Hall, Upper Saddle River (1982)
30. Endou, T., Zhao, Q.: Generation of comprehensible decision trees through evolution of training data. In: Proceedings of the 2002 Congress on Evolutionary Computation (CEC'02), vol. 2, pp. 1221–1225. IEEE, Upper Saddle River (2002)
31. Frédéric, R., Serge, G.: An efficient nearest neighbor classifier. In: Hybrid Evolutionary Algorithms, pp. 127–145. Springer, Berlin (2007)
32. Gan, J.Q., Hasan, B.A.S., Tsui, C.S.L.: A filter-dominating hybrid sequential forward floating search method for feature subset selection in high-dimensional space. Int. J. Mach. Learn. Cybern. **5**(3), 413–423 (2014)
33. Gao, W., Hu, L., Zhang, P., Wang, F.: Feature selection by integrating two groups of feature evaluation criteria. Expert Syst. Appl. **110**, 11–19 (2018)
34. García-Osorio, C., de Haro-García, A., García-Pedrajas, N.: Democratic instance selection: a linear complexity instance selection algorithm based on classifier ensemble concepts. Artif. Intell. **174**(5–6), 410–441 (2010)
35. García-Pedrajas, N.: Evolutionary computation for training set selection. Wiley Interdisciplinary Reviews. Data Min. Knowl. Discov. **1**(6), 512–523 (2011)
36. GarcíA-Pedrajas, N., De Haro-GarcíA, A., PéRez-RodríGuez, J.: A scalable approach to simultaneous evolutionary instance and feature selection. Inf. Sci. **228**, 150–174 (2013)
37. García-Pedrajas, N., de Haro-García, A., Pérez-Rodríguez, J.: A scalable memetic algorithm for simultaneous instance and feature selection. Evol. Comput. **22**(1), 1–45 (2014)
38. Gates, G.: The reduced nearest neighbor rule (corresp.). IEEE Trans. Inf. Theory **18**(3), 431–433 (1972)
39. Ghaemi, M., Feizi-Derakhshi, M.R.: Forest optimization algorithm. Expert Syst. Appl. **41**(15), 6676–6687 (2014)
40. Ghaemi, M., Feizi-Derakhshi, M.R.: Feature selection using forest optimization algorithm. Pattern Recognit. **60**, 121–129 (2016)

41. Ghareb, A.S., Bakar, A.A., Hamdan, A.R.: Hybrid feature selection based on enhanced genetic algorithm for text categorization. Expert Syst. Appl. **49**, 31–47 (2016)
42. Grochowski, M., Jankowski, N.: Comparison of instance selection algorithms ii. results and comments. In: International Conference on Artificial Intelligence and Soft Computing, pp. 580–585. Springer, Berlin (2004)
43. Gutlein, M., Frank, E., Hall, M., Karwath, A.: Large-scale attribute selection using wrappers. In: IEEE Symposium on Computational Intelligence and Data Mining, CIDM'09, pp. 332–339. IEEE, Piscataway (2009)
44. Guyon, I., Elisseeff, A.: An introduction to variable and feature selection. J. Mach. Learn. Res. **3**, 1157–1182 (2003)
45. Guyon, I., Elisseeff, A.: An introduction to feature extraction. In: Feature Extraction, pp. 1–25. Springer, Berlin (2006)
46. Hall, M.A.: Correlation-based feature selection of discrete and numeric class machine learning. In: Proceedings of the Seventeenth International Conference on Machine Learning (2000)
47. Hall, M.A., Holmes, G.: Benchmarking attribute selection techniques for discrete class data mining. IEEE Trans. Knowl. Data Eng. **15**(6), 1437–1447 (2003)
48. Hart, P.: The condensed nearest neighbor rule (corresp.). IEEE Trans. Inf. Theory **14**(3), 515–516 (1968)
49. Ho, S.Y., Liu, C.C., Liu, S.: Design of an optimal nearest neighbor classifier using an intelligent genetic algorithm. Pattern Recognit. Lett. **23**(13), 1495–1503 (2002)
50. Hsu, H.H., Hsieh, C.W., Lu, M.D.: Hybrid feature selection by combining filters and wrappers. Expert Syst. Appl. **38**(7), 8144–8150 (2011)
51. John, G.H., Kohavi, R., Pfleger, K.: Irrelevant features and the subset selection problem. In: Machine Learning Proceedings 1994, pp. 121–129. Elsevier, Amsterdam (1994)
52. Kabir, M.M., Shahjahan, M., Murase, K.: A new local search based hybrid genetic algorithm for feature selection. Neurocomputing **74**(17), 2914–2928 (2011)
53. Kim, S.W., Oommen, B.J.: Enhancing prototype reduction schemes with recursion: a method applicable for "large" data sets. IEEE Trans. Syst. Man Cybern. B **34**(3), 1384–1397 (2004)
54. Kira, K., Rendell, L.A.: A practical approach to feature selection. In: Machine Learning Proceedings 1992, pp. 249–256. Elsevier, Amsterdam (1992)
55. Kohavi, R., John, G.H.: Wrappers for feature subset selection. Artif. Intell. **97**(1–2), 273–324 (1997)
56. Koller, D., Sahami, M.: Toward optimal feature selection. Tech. Rep., Stanford InfoLab (1996)
57. Kononenko, I.: Estimating attributes: analysis and extensions of relief. In: European Conference on Machine Learning, pp. 171–182. Springer, Berlin (1994)
58. Kononenko, I., Šimec, E., Robnik-Šikonja, M.: Overcoming the myopia of inductive learning algorithms with ReliefF. Appl. Intell. **7**(1), 39–55 (1997)
59. Kordos, M., Blachnik, M.: Instance selection with neural networks for regression problems. In: International Conference on Artificial Neural Networks, pp. 263–270. Springer, Berlin (2012)
60. Kuncheva, L.I., Bezdek, J.C.: Nearest prototype classification: clustering, genetic algorithms, or random search? IEEE Trans. Syst. Man, Cybern. C **28**(1), 160–164 (1998)
61. Kuncheva, L.I., Jain, L.C.: Nearest neighbor classifier: simultaneous editing and feature selection. Pattern Recognit. Lett. **20**(11–13), 1149–1156 (1999)
62. Leyva, E., González, A., Pérez, R.: Three new instance selection methods based on local sets: a comparative study with several approaches from a bi-objective perspective. Pattern Recognit. **48**(4), 1523–1537 (2015)
63. Liu, H., Motoda, H.: Feature Selection for Knowledge Discovery and Data Mining, vol. 454. Springer Science & Business Media, New York (2012)
64. Liu, H., Yu, L.: Toward integrating feature selection algorithms for classification and clustering. IEEE Trans. Knowl. Data Eng. **17**(4), 491–502 (2005)

65. Maihami, V., Yaghmaee, F.: A genetic-based prototyping for automatic image annotation. Comput. Electr. Eng. **70**, 400–412 (2018)
66. Mirjalili, S., Saremi, S., Mirjalili, S.M., Coelho, L.D.S.: Multi-objective grey wolf optimizer: a novel algorithm for multi-criterion optimization. Expert Syst. Appl. **47**, 106–119 (2016)
67. Murillo, J., Guillaume, S., Spetale, F., Tapia, E., Bulacio, P.: Set characterization-selection towards classification based on interaction index. Fuzzy Sets Syst. **270**, 74–89 (2015)
68. Nakariyakul, S.: High-dimensional hybrid feature selection using interaction information-guided search. Knowl.-Based Syst. **145**, 59–66 (2018)
69. Nakariyakul, S., Liu, Z.P., Chen, L.: A sequence-based computational approach to predicting PDZ domain-peptide interactions. Biochimica et Biophysica Acta (BBA)-Proteins Proteomics **1844**(1), 165–170 (2014)
70. Novaković, J.: Toward optimal feature selection using ranking methods and classification algorithms. Yugoslav J. Oper. Res. **21**(1). 119–135 (2016)
71. Oates, T., Jensen, D.: The effects of training set size on decision tree complexity. In: Proceedings of 14th International Conference on Machine Learning. Citeseer (1997)
72. Oh, I.S., Lee, J.S., Moon, B.R.: Hybrid genetic algorithms for feature selection. IEEE Trans. Pattern Anal. Mach. Intell. **26**(11), 1424–1437 (2004)
73. Olvera-López, J.A., Carrasco-Ochoa, J.A., Martínez-Trinidad, J.F.: Sequential search for decremental edition. In: International Conference on Intelligent Data Engineering and Automated Learning, pp. 280–285. Springer, Berlin (2005)
74. Olvera-López, J.A., Martínez-Trinidad, J.F., Carrasco-Ochoa, J.A.: Restricted sequential floating search applied to object selection. In: International Workshop on Machine Learning and Data Mining in Pattern Recognition, pp. 694–702. Springer, Berlin (2007)
75. Olvera-López, J.A., Carrasco-Ochoa, J., Kittler, J., et al.: Prototype selection based on sequential search. Intell. Data Anal. **13**(4), 599–631 (2009)
76. Olvera-López, J.A., Carrasco-Ochoa, J.A., Martínez-Trinidad, J.F.: A new fast prototype selection method based on clustering. Pattern Anal. Appl. **13**(2), 131–141 (2010)
77. Olvera-López, J.A., Carrasco-Ochoa, J.A., Martínez-Trinidad, J.F., Kittler, J.: A review of instance selection methods. Artif. Intell. Rev. **34**(2), 133–143 (2010)
78. Peng, H., Long, F., Ding, C.: Feature selection based on mutual information criteria of max-dependency, max-relevance, and min-redundancy. IEEE Trans. Pattern Anal. Mach. Intell. **27**(8), 1226–1238 (2005)
79. Pérez-Rodríguez, J., Arroyo-Peña, A.G., García-Pedrajas, N.: Simultaneous instance and feature selection and weighting using evolutionary computation: proposal and study. Appl. Soft Comput. **37**, 416–443 (2015)
80. Picek, S., Golub, M., Jakobovic, D.: Evaluation of crossover operator performance in genetic algorithms with binary representation. In: International Conference on Intelligent Computing, pp. 223–230. Springer, Berlin (2011)
81. Price, K., Storn, R.M., Lampinen, J.A.: Differential evolution: a practical approach to global optimization. Springer Science & Business Media, New York (2006)
82. Pudil, P., Novovičová, J., Kittler, J.: Floating search methods in feature selection. Pattern recognition letters **15**(11), 1119–1125 (1994)
83. Ramirez-Cruz, J.F., Fuentes, O., Alarcon-Aquino, V., Garcia-Banuelos, L.: Instance selection and feature weighting using evolutionary algorithms. In: 15th International Conference on Computing, CIC'06, pp. 73–79. IEEE, Piscataway (2006)
84. Ritter, G., Woodruff, H., Lowry, S., Isenhour, T.: An algorithm for a selective nearest neighbor decision rule (corresp.). IEEE Trans. Inf. Theory **21**(6), 665–669 (1975)
85. Ros, F., Guillaume, S., Pintore, M., Chrétien, J.R.: Hybrid genetic algorithm for dual selection. Pattern Anal. Appl. **11**(2), 179–198 (2008)
86. Ruggieri, S.: Efficient c4. 5 [classification algorithm]. IEEE Trans. Knowl. Data Eng. **14**(2), 438–444 (2002)
87. Ruiz, R., Riquelme, J.C., Aguilar-Ruiz, J.S., García-Torres, M.: Fast feature selection aimed at high-dimensional data via hybrid-sequential-ranked searches. Expert Syst. Appl. **39**(12), 11094–11102 (2012)

88. Schmidhuber, J.: Deep learning in neural networks: an overview. Neural Netw. **61**, 85–117 (2015)
89. Sebban, M., Nock, R.: Instance pruning as an information preserving problem. In: Proceedings of the Seventeenth International Conference on Machine Learning, pp. 855–862. Morgan Kaufmann, Burlington (2000)
90. Skalak, D.B.: Prototype and feature selection by sampling and random mutation hill climbing algorithms. In: Machine Learning Proceedings 1994, pp. 293–301. Elsevier, Amsterdam (1994)
91. Skalak, D.B.: Prototype selection for composite nearest neighbor classifiers. Ph.D. Thesis, University of Massachusetts at Amherst (1997)
92. Smith, D.R.: The design of divide and conquer algorithms. Sci. Comput. Program. **5**, 37–58 (1985)
93. Song, Y., Liang, J., Lu, J., Zhao, X.: An efficient instance selection algorithm for k nearest neighbor regression. Neurocomputing **251**, 26–34 (2017)
94. Tibshirani, R.: The lasso method for variable selection in the cox model. Stat. Med. **16**(4), 385–395 (1997)
95. Tsai, C.F., Eberle, W., Chu, C.Y.: Genetic algorithms in feature and instance selection. Knowl.-Based Syst. **39**, 240–247 (2013)
96. Wang, L., Wang, Y., Chang, Q.: Feature selection methods for big data bioinformatics: a survey from the search perspective. Methods **111**, 21–31 (2016)
97. Wettschereck, D., Aha, D.W., Mohri, T.: A review and empirical evaluation of feature weighting methods for a class of lazy learning algorithms. Artif. Intell. Rev. **11**(1–5), 273–314 (1997)
98. Wilson, D.L.: Asymptotic properties of nearest neighbor rules using edited data. IEEE Trans. Syst. Man Cybern. SMC-2(3), 408–421 (1972)
99. Wilson, D.R., Martinez, T.R.: Reduction techniques for instance-based learning algorithms. Mach. Learn. **38**(3), 257–286 (2000)
100. Yang, Y., Webb, G.I., Wu, X.: Discretization methods. In: Data Mining and Knowledge Discovery Handbook, pp. 101–116. Springer, Berlin (2009)
101. Yu, L., Liu, H.: Feature selection for high-dimensional data: a fast correlation-based filter solution. In: Proceedings of the 20th International Conference on Machine Learning (ICML-03), pp. 856–863 (2003)
102. Yu, L., Liu, H.: Efficient feature selection via analysis of relevance and redundancy. J. Mach. Learn. Res. **5**, 1205–1224 (2004)
103. Yuan, L., Liu, J., Tang, X.: Combining example selection with instance selection to speed up multiple-instance learning. Neurocomputing **129**, 504–515 (2014)
104. Zawbaa, H.M., Emary, E., Grosan, C., Snasel, V.: Large-dimensionality small-instance set feature selection: a hybrid bio-inspired heuristic approach. Swarm Evol. Comput. **42**, 29–42 (2018)
105. Zhou, P., Hu, X., Li, P., Wu, X.: OFS-density: a novel online streaming feature selection method. Pattern Recognit. **86**, 48–61 (2019)
106. Zhu, X., Wu, X.: Scalable representative instance selection and ranking. In: 18th International Conference on Pattern Recognition, ICPR 2006, vol. 3, pp. 352–355. IEEE, Piscataway (2006)

Chapter 5
Approximating Spectral Clustering via Sampling: A Review

Nicolas Tremblay and Andreas Loukas

5.1 Introduction

Clustering is a cornerstone of our learning process and, thus, of our understanding of the world. Indeed, we can all distinguish between a rose and a tulip precisely because we have learned what these flowers *are*. Plato would say that we learned the Idea—or Form [120]—of both the rose and the tulip, which then enables us to recognize all instances of such flowers. A machine learner would say that we learned two *classes*: their most discriminating features (shape, size, number of petals, smell, etc.) as well as their possible intra-class variability.

Mathematically speaking, the first step on the road to classifying objects (such as flowers) is to create an abstract representation of these objects: with each object i we associate a feature vector $\mathbf{p}_i \in \mathbb{R}^d$, where the dimension d of the vector corresponds to the number of features one chooses to select for the classification task. The space $\mathbb{R}^d$ in this context is sometimes called the *feature space*. The choice of representation will obviously have a strong impact on the subsequent classification performance. Say that in the flower example we choose to represent each flower by only $d = 3$ features: the average color of each RGB channel (level of red, green, and blue) of its petals. This choice is not fail-proof: even though the archetype of the rose is red and the archetype of the tulip is yellow, we know that some varieties of both flowers can have very similar colors and thus a classification solely based on the color will necessarily lead to confusion. In fact, there are many

N. Tremblay (✉)
CNRS, University of Grenoble Alpes, GIPSA-Lab, Saint-Martin-d'Hères, France
e-mail: nicolas.tremblay@grenoble-inp.fr

A. Loukas
Ecole Polytechnique Fédérale de Lausanne, Lausanne, Switzerland
e-mail: andreas.loukas@epfl.ch

F. Ros, S. Guillaume (eds.), *Sampling Techniques for Supervised or Unsupervised Tasks*, Unsupervised and Semi-Supervised Learning,
https://doi.org/10.1007/978-3-030-29349-9_5

different ways of choosing features: from features based on the expert knowledge of a botanist to features learned by a deep learning architecture from many instances of labeled images of roses and tulips, via features obtained by hybrid methods more-or-less based on human intelligence (such as the first few components of a principal component analysis of expert-based features).

The second step on the road to classifying n objects is to choose a machine learning algorithm that groups the set of n points $\mathsf{P} = (\mathbf{p}_1, \ldots, \mathbf{p}_n)$ in k classes (k may be known in advance or determined by the algorithm itself). Choosing an appropriate algorithm depends on the context:

- **Availability of pre-labeled data.** Classifying the points P in k classes may be seen as assigning a label (such as "rose" or "tulip" in our $k = 2$ example) to each of the points. If one has access to some pre-labeled data, we are in the case of *supervised learning*: a more-or-less parametrized model is first learned from the pre-labeled data and then applied to the unlabeled points that need classification. If one does not have access to any pre-labeled data, we are in the case of *unsupervised learning* where classes are typically inferred only via geometrical consideration of the distribution of points in the feature space. If one has only access to a few labeled data, we are in the in-between case of *semi-supervised learning* where the known labels are typically propagated in one form or another in the feature space.
- **Inductive vs transductive learning.** Another important characteristic of a classification algorithm is whether it can be used to classify only the set of points P at hand (transductive), or if it can also be directly used to classify any never-seen data point $\mathbf{p}_{n+1}$ (inductive).

This chapter focuses on the family of algorithms jointly referred to as *spectral clustering*. These algorithms are unsupervised and transductive: no label is known in advance and one may not naturally[1] extend the results obtained on P to never-seen data points. Another particularity of spectral clustering algorithms is that the number of classes k is known in advance.

Spectral clustering algorithms have received a large attention in the last two decades due to their good performance on a wide range of different datasets, as well as their ease of implementation. In a nutshell, they combine three steps:

1. **Graph construction.** A sparse similarity graph is built between the n points.
2. **Spectral embedding.** The first k eigenvectors of a graph representative matrix (such as the Laplacian) are computed.
3. **Clustering.** k-means is performed on these spectral features, to obtain k clusters.

For background information about spectral clustering, such as several justifications of its performance, out-of-sample extensions, as well as comparisons with local methods, the interested reader is referred to the recent book chapter [144].

[1] Out-of-sample extensions of spectral clustering do exist (see, for instance, Section 5.3.6 of [144]), but they require additional work.

One of the drawbacks of spectral clustering is its computational cost as n, d, and/or k become large (see Sect. 5.2.3 for a discussion on the cost). Since the turn of the century, a large number of authors have striven to reduce the computational cost while keeping the high level of classification performance. The majority of such accelerating methods are based on sampling: they reduce the dimension of a sub-problem of spectral clustering, compute a low-cost solution in small dimension, and lift the result back to the original space.

The goal of this chapter is to review existing sampling methods for spectral clustering, focusing especially on their approximation guarantees. Some of the fundamental questions we are interested in are: *where is the sampling performed and what is sampled precisely? how should the reduced approximate solutions be lifted back to the original space? what is the computational gain? what is the control on performances—if it exists?* Given the breadth of the literature on the subject, we do not try to be exhaustive, but rather to illustrate the key ways that sampling can be used to provide acceleration, paying special attention on recent developments on the subject.

Chapter Organization We begin by recalling in Sect. 5.2 the prototypical spectral clustering algorithm. We also provide some intuitive and formal justification of why it works. The next three sections classify the different methods of the literature depending on where the sampling is performed with respect to the three steps of spectral clustering:

- Section 5.3 details methods that sample directly in the original feature space.
- Section 5.4 assumes that the similarity graph is given and details methods that sample nodes and/or edges to approximate the spectral embedding.
- Section 5.5 assumes that the spectral embedding is given and details methods to accelerate the k-means step.

Finally, Sect. 5.6 gives perspective on the limitations of existing works and discusses key open problems.

Notation Scalars, such as λ or d, are written with lowercase letters. Vectors, such as $\mathbf{u}$, $\mathbf{z}$, or the all-one vector $\mathbf{1}$, are denoted by lowercase bold letters. Matrices, such as $\mathbf{W}$ or $\mathbf{L}$, are denoted by bold capital letters. Ensembles are denoted by serif font capital letters, such as C or X. The "tilde" will denote approximations, such as in $\tilde{\mathbf{z}}$ or $\tilde{\mathbf{U}}_k$. We use so-called Matlab notations to slice matrices: given a set of indices S of size m and an $n \times n$ matrix $\mathbf{W}$, $\mathbf{W}(\mathsf{S}, :) \in \mathbb{R}^{m\times n}$ is $\mathbf{W}$ reduced to the lines indexed by S, $\mathbf{W}(:, \mathsf{S}) \in \mathbb{R}^{n\times m}$ is $\mathbf{W}$ reduced to the columns indexed by S, and $\mathbf{W}(\mathsf{S}, \mathsf{S}) \in \mathbb{R}^{m\times m}$ is $\mathbf{W}$ reduced to the lines and columns indexed by S. The equation $\mathbf{U}_k = \mathbf{U}(:, :k)$ defines $\mathbf{U}_k$ as the reduction of $\mathbf{U}$ to its first k columns. Also, $\mathbf{C}^\top$ is the transpose of matrix $\mathbf{C}$ and $\mathbf{C}^+$ its Moore–Penrose pseudo-inverse. The operator $\mathbf{X} = \mathrm{diag}(\mathbf{x})$ takes as an input a vector $\mathbf{x} \in \mathbb{R}^n$ and returns an $n \times n$ diagonal matrix $\mathbf{X}$ featuring $\mathbf{x}$ in its main diagonal, i.e., $\mathbf{X}(i, j) = \mathbf{x}(i)$ if $i = j$ and $\mathbf{X}(i, j) = 0$, otherwise. Finally, we will consider graphs in a large part of this paper. We will denote by $G = (\mathsf{V}, \mathsf{E}, \mathbf{W})$ the undirected weighted graph of $|\mathsf{V}| = n$ nodes interconnected by $|\mathsf{E}| = e$ edges. $e_{ij} \in \mathsf{E}$ is the edge connecting nodes v_i and v_j, with weight $\mathbf{W}(i, j) \geq 0$. Matrix $\mathbf{W}$ is the adjacency matrix of G. As G is considered undirected, $\mathbf{W}$ is also symmetric. In general, $\mathbf{W}$ can be any symmetric matrix with positive entries, but we usually prefer to work with sparse graphs without self-loops, in which case the matrix is also sparse and has a zero diagonal.

5.2 Spectral Clustering

The input of spectral clustering algorithms consists of (1) a set of points $\mathsf{P} = (\mathbf{p}_1, \mathbf{p}_2, \ldots, \mathbf{p}_n)$ (also called featured vectors) representing n objects in a feature space of dimension d, and (2) the number of classes k in which to classify these objects. The output is a partition of the n objects in k disjoint clusters. The prototypical spectral clustering algorithm [102, 121] dates back in fact to fundamental ideas by Fiedler [46] and entails the following steps:

Algorithm 1. The prototypical spectral clustering algorithm
Input. A set of n points $\mathsf{P} = (\mathbf{p}_1, \mathbf{p}_2, \ldots, \mathbf{p}_n)$ in dimension d and a number of desired clusters k.

1. Graph construction (optional)
   (a) Compute the kernel matrix $\mathbf{K} \in \mathbb{R}^{n \times n}$: $\forall(i, j)$, $\mathbf{K}(i, j) = \kappa(\|\mathbf{p}_i - \mathbf{p}_j\|_2)$.
   (b) Compute $\mathbf{W} = s(\mathbf{K})$, a sparsified version of $\mathbf{K}$.
   (c) Interpret $\mathbf{W}$ as the adjacency matrix of a weighted undirected graph G.
2. Spectral embedding
   (a) Compute the eigenvectors $\mathbf{u}_1, \mathbf{u}_2, \cdots, \mathbf{u}_k$ associated with the k smallest eigenvalues of a graph representative matrix $\mathbf{R}$ (usually a Laplacian) computed from $\mathbf{W}$.
   (b) Set $\mathbf{U}_k = [\,\mathbf{u}_1 | \,\mathbf{u}_2 | \cdots | \,\mathbf{u}_k\,] \in \mathbb{R}^{n \times k}$.
   (c) Embed the i-th node to $\mathbf{x}_i = \frac{\mathbf{U}_k(i,:)^\top}{q(\|\mathbf{U}_k(i,:)\|_2)}$, with $q(\cdot)$ a normalizing function.
3. Clustering
   (a) Use k-means on $\mathbf{x}_1, \ldots, \mathbf{x}_n$ in order to identify k centroids $\mathbf{c}_1, \ldots, \mathbf{c}_k$.
   (b) Voronoi tessellation: construct one cluster per centroid $\mathbf{c}_\ell$ and assign each object i to the cluster of the centroid closest to $\mathbf{x}_i$.

Output: A partition of the n points in k clusters.

A few comments are in order:

- A common choice of kernel in step 1a is the radial basis function (RBF) kernel $\kappa(\|\mathbf{p}_i - \mathbf{p}_j\|_2) = \exp\left(-\|\mathbf{p}_i - \mathbf{p}_j\|_2^2/\sigma^2\right)$ for some user-defined σ. The sparsification s of $\mathbf{K}$ usually entails setting the diagonal to 0 and keeping only the k largest entries of each column (i.e., set all others to 0). The obtained matrix $\mathbf{K}_{\text{sp}}$ is not symmetric in general and a final "symmetrization" step $\mathbf{W} = \mathbf{K}_{\text{sp}} + \mathbf{K}_{\text{sp}}^{\top}$ is necessary to obtain a matrix $\mathbf{W}$ interpretable as the adjacency matrix of a weighted undirected graph[2] $G = (\mathsf{V}, \mathsf{E}, \mathbf{W})$. This graph is called the k nearest neighbor (k-NN) similarity graph (note that the k used in this paragraph has nothing to do with the number of clusters). Other kernel functions κ and sparsification methods are possible (see Section 2 of [138] for instance).
- There are several possibilities for choosing the graph representative matrix $\mathbf{R}$ in step 2a. We consider three main choices [138]. Let us denote by $\mathbf{D}$ the diagonal degree matrix such that $\mathbf{D}(i, i) = \sum_j \mathbf{W}(i, j)$ is the (weighted) degree of node v_i. We define the *combinatorial* graph Laplacian matrix $\mathbf{L} = \mathbf{D} - \mathbf{W}$, the *normalized* graph Laplacian matrix $\mathbf{L}_n = \mathbf{I} - \mathbf{D}^{-1/2}\mathbf{W}\mathbf{D}^{-1/2}$, and the *random walk* Laplacian $\mathbf{L}_{rw} = \mathbf{I} - \mathbf{D}^{-1}\mathbf{W}$. Other popular choices include[3] the non-backtracking matrix [73], degree-corrected versions of the modularity matrix [2], the Bethe–Hessian matrix [114], or similar deformed Laplacians [34].
- The normalizing function $q(\cdot)$ used in step 2c depends on which representative matrix is chosen. In the case of the Laplacians, experimental evidence as well as some theoretical arguments [138] support using a unit norm normalization for the eigenvectors of $\mathbf{L}_n$ (i.e., q is the identity function), and no normalization for the eigenvectors of $\mathbf{L}$ and $\mathbf{L}_{rw}$ (i.e., $q(\cdot) = 1$).
- Step 1 of the algorithm is "optional" in the sense that in some cases the input is not a set of points but directly a graph. For instance, it could be a graph representing a social network between n individuals, where each node is an individual and there is an edge between two nodes if they know each other. The weight on each edge can represent the strength of their relation (for instance, close to 0 if they barely know each other, and close to 1 if they are best friends). The goal is then to classify individuals based on the structure of these social connections and is usually referred to as *community detection* in this context [47]. Given the input graph, and the number k of communities to identify, one can run spectral algorithms starting directly at step 2. Readers only interested in such applications can skip Sect. 5.3, which is devoted to sampling techniques designed to accelerate step 1.

---

[2]Each node v_i of V represents a point $\mathbf{p}_i$, an undirected edge exists between nodes v_i and v_j if and only if $\mathbf{W}(i, j) \neq 0$, and the weight of that connection is $\mathbf{W}(i, j)$.

[3]In some of these examples, the k largest eigenvalues (instead of the k lowest in the Laplacian cases) of the representative matrix, and especially their corresponding eigenvectors, are of interest. This is only a matter of sign of the matrix $\mathbf{R}$ and has no impact on the general discussion.

After the spectral embedding $\mathsf{X} = (\mathbf{x}_1, \ldots, \mathbf{x}_n)$ has been identified, spectral clustering uses k-means in order to find the set of k centroids $\mathsf{C} = (\mathbf{c}_1, \ldots, \mathbf{c}_k)$ that best represents the data. Formally, the k-means cost function to minimize reads:

$$f(\mathsf{C}; \mathsf{X}) = \sum_{\mathbf{x} \in \mathsf{X}} \min_{\mathbf{c} \in \mathsf{C}} \|\mathbf{x} - \mathbf{c}\|_2^2. \tag{5.1}$$

We would ideally hope to identify the set of k centroids C^* minimizing $f(\mathsf{C}; \mathsf{X})$. Solving exactly this problem is NP-hard [42], so one commonly resorts to approximation and heuristic solutions (see, for instance, [128] for details on different such heuristics). The most famous is the so-called Lloyd-Max heuristic algorithm:

Algorithm 2. The Lloyd-Max algorithm [87]
Input. Set of n points $\mathsf{X} = (\mathbf{x}_1, \mathbf{x}_2, \ldots, \mathbf{x}_n)$ and number of desired clusters k.

1. Start from an initial guess C_{ini} of k centroids
2. Iterate until convergence:
   (a) Assign each point $\mathbf{x}_i$ to its closest centroid to obtain a partition of X in k clusters.
   (b) Move each centroid $\mathbf{c}_\ell$ to the average position of all points in cluster ℓ.

Output: A set of k centroids $\mathsf{C} = (\mathbf{c}_1, \ldots, \mathbf{c}_k)$.

When the clusters are sufficiently separated and C_{ini} is not too far from the optimal centroids, then the Lloyd-Max algorithm converges to the correct solution [75]. Otherwise, it typically ends up in a local minimum.

A Remark on Notation Two quantities of fundamental importance in spectral clustering are the eigenvalues λ_i and especially the eigenvectors $\mathbf{u}_i$ of the graph Laplacian matrix. We adopt the graph theoretic convention of sorting eigenvalues in non-decreasing order: $0 = \lambda_1 \leq \lambda_2 \leq \ldots \leq \lambda_n$. Also, for reasons of brevity, we overload notation and use the same symbol for the spectrum of the three Laplacians $\mathbf{L}$, $\mathbf{L}_n$, and $\mathbf{L}_{rw}$. Thus, we advise the reader to rely on the context in order to discern which Laplacian gives rise to the eigenvalues and eigenvectors. Finally, the reader should keep in mind that the largest eigenvalue is always bounded by 2 for $\mathbf{L}_n$ and $\mathbf{L}_{rw}$.

5.2.1 *An Illustration of Spectral Clustering*

The first two steps of the algorithm can be understood as a non-linear transformation from the initial feature space to another feature space (that we call spectral feature space or spectral embedding): a transformation of features $\mathbf{p}_i$ in $\mathbb{R}^d$ to spectral features $\mathbf{x}_i$ in $\mathbb{R}^k$. The first natural question that arises is why do we run k-means on the spectral features $\mathsf{X} = (x_1, \ldots, \mathbf{x}_n)$ that are subject to parameter tuning and costly to compute, rather than directly run k-means on the original P? Figures 5.1 and 5.2 illustrate the answer.

In Fig. 5.1, we show the result of k-means directly on a set of artificial features P known as the two half-moons dataset. In this example, the intuitive ground truth is that each half-moon corresponds to a class that we want to recover. Running k-means directly in this 2D feature space will necessarily output a linear separation between the two obtained Voronoi cells and will thus necessarily fail, as no straight line can separate the two half-moons.

Spectral clustering, via the computation of the spectral features of a similarity graph, transforms these original features P in spectral features X that are typically linearly separable by k-means: the two half-moons are successfully recovered! We illustrate this in Fig. 5.2. In the next section, we will examine a theoretical argument aiming to justify this phenomenon.

5.2.2 *Justification of Spectral Clustering*

A popular approach—and by no means the only one, see Sect. 5.2.2.3—to justify spectral clustering algorithms stems from its connection to graph partitioning. Suppose that the similarity graph $G = (\mathsf{V}, \mathsf{E}, \mathbf{W})$ has been obtained and we want to compute a partition[4] $\mathscr{P} = \{\mathsf{V}_1, \mathsf{V}_2, \ldots, \mathsf{V}_k\}$ of the nodes V in k groups. Intuitively, a good clustering objective function should favor strongly connected nodes to end

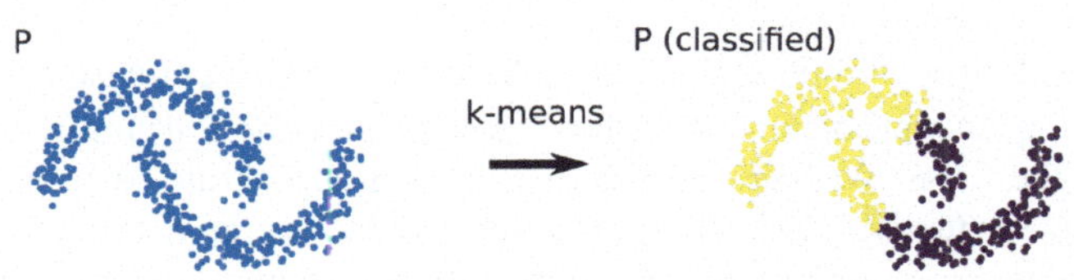

Fig. 5.1 Left: the two half-moons synthetic dataset ($n = 500, d = 2, k = 2$). Right: k-means with $k = 2$ directly on P is unsuccessful to separate the two half-moons

[4]By definition, a *partition* $\mathscr{P} = \{\mathsf{V}_1, \mathsf{V}_2, \ldots, \mathsf{V}_k\}$ of the nodes V is such that $\cup_{\ell=1,\ldots,k}\mathsf{V}_\ell = \mathsf{V}$ and $\forall \ell \neq \ell', \mathsf{V}_\ell \cap \mathsf{V}_{\ell'} = \emptyset$.

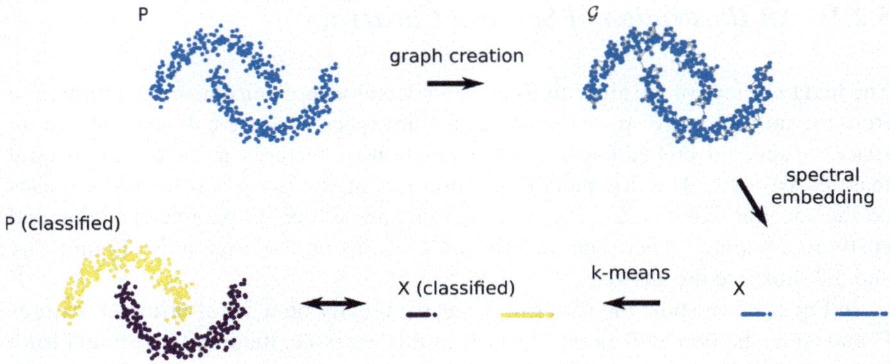

Fig. 5.2 Illustration of the spectral clustering algorithm on the two half-moons dataset ($n = 500$, $d = 2$, $k = 2$). The graph is created with a RBF kernel and via a sparsification done with k-nearest neighbors (with $k = 5$). The spectral embedding is done with the two eigenvectors associated with the two smallest eigenvalues of the combinatorial Laplacian matrix $\mathbf{L}$. The embedding X is here in practice in 1$\mathbf{D}$ as the first eigenvector of $\mathbf{L}$ is always constant and thus not discriminative (to confirm this, first show that $\mathbf{L}$ is a PSD matrix and then prove that $\mathbf{Lc} = 0$ for any constant vector $\mathbf{c}$). Observe how the two clusters are now linearly separable in the spectral feature space. k-means on *these* features successfully recovers the two half-moons

up in the same subset, and nodes that are far apart in the graph to end up in different subsets. This intuition can be formalized with graph cuts.

Considering two groups V_1 and V_2, define $w(\mathsf{V}_1, \mathsf{V}_2) = \sum_{i\in\mathsf{V}_1}\sum_{j\in\mathsf{V}_2} \mathbf{W}(i, j)$ to be the total weight of all links connecting V_1 to V_2. Also, denote by $\bar{\mathsf{V}}_\ell$ the complement of V_ℓ in V, such that $\frac{1}{2}w(\mathsf{V}_\ell, \bar{\mathsf{V}}_\ell)$ is the total weight one needs to cut in order to disconnect V_ℓ from the rest of the graph. Given these definitions, the simplest graph cut objective function, denoted by `cut`, is:

$$\texttt{cut}\left(\mathscr{P} = \{\mathsf{V}_1, \ldots, \mathsf{V}_k\}\right) = \frac{1}{2}\sum_{\ell=1}^{k} w\left(\mathsf{V}_\ell, \bar{\mathsf{V}}_\ell\right). \tag{5.2}$$

The best partition according to the `cut` criterion is $\mathscr{P}^* = \operatorname{argmin}_{\mathscr{P}} \texttt{cut}(\mathscr{P})$. For $k = 2$, solving this problem can be done exactly in $O(ne + n^2\log(n))$ amortized time using the Stoer–Wagner algorithm [126] and approximated in nearly linear time [68]. Nevertheless, this criterion is not satisfactory as it often separates an individual node from the rest of the graph, with no attention to the balance of the sizes or volumes of the groups. In clustering, one usually wants to partition into groups that are "large enough." There are two famous ways to balance the previous cost in the machine learning literature[5]: the *ratio cut* [143] and *normalized cut* [121] cost functions, respectively defined as:

[5]The reader should note that in the graph theory literature, the measure of conductance is preferred over `ncut`. Conductance is $\max_\ell w(\mathsf{V}_\ell, \bar{\mathsf{V}}_\ell)/w(\mathsf{V}_\ell)$. The two measures are equivalent when $k = 2$.

$$\texttt{rcut}(\mathscr{P}) = \frac{1}{2}\sum_{\ell=1}^{k}\frac{w(\mathsf{V}_\ell, \bar{\mathsf{V}}_\ell)}{|\mathsf{V}_\ell|} \quad \text{and} \quad \texttt{ncut}(\mathscr{P}) = \frac{1}{2}\sum_{\ell=1}^{k}\frac{w(\mathsf{V}_\ell, \bar{\mathsf{V}}_\ell)}{\text{vol}(\mathsf{V}_\ell)}, \tag{5.3}$$

where $|\mathsf{V}_\ell|$ is the number of nodes in V_ℓ and $\text{vol}(\mathsf{V}_\ell) = \sum_{i\in\mathsf{V}_\ell}\sum_{j\in\mathsf{V}}\mathbf{W}(i, j)$ is the so-called volume of V_ℓ. The difference between them is that ncut favors clusters of large volume, whereas rcut only considers cluster size—though for a d-regular graph with unit weights the two measures match (up to multiplication by $1/d$). Unfortunately, it is hard to minimize these cost functions directly: minimizing these two balanced costs is NP-hard [121, 139] and one needs to search over the space of all possible partitions which is of exponential size.

A Continuous Relaxation Spectral clustering may be interpreted as a continuous relaxation of the above minimization problems. Without loss of generality, in the following we concentrate on relaxing the rcut minimization problem (ncut is relaxed almost identically). Given a partition $\mathscr{P} = (\mathsf{V}_1, \ldots, \mathsf{V}_k)$, let us define

$$\mathbf{C} = \left(\frac{\mathbf{z}_1}{\sqrt{|\mathsf{V}_1|}}|\ldots|\frac{\mathbf{z}_k}{\sqrt{|\mathsf{V}_k|}}\right) \in \mathbb{R}^{n\times k}, \tag{5.4}$$

where $\mathbf{z}_\ell \in \mathbb{R}^n$ is the indicator vector of V_ℓ:

$$\mathbf{z}_\ell(i) = \begin{cases} 1 & \text{if node } i \in \mathsf{V}_\ell, \\ 0 & \text{otherwise.} \end{cases} \tag{5.5}$$

It will prove useful in the following to remark that, independently of how the partitions are chosen, we always have that $\mathbf{C}^\top\mathbf{C} = \mathbf{I}$, the identity matrix in dimension k. With this in place, the problem of minimizing rcut can be rewritten as (see discussion in [138]):

$$\min_{\mathbf{C}\in\mathbb{R}^{n\times k}} \text{tr}\left(\mathbf{C}^\top\mathbf{L}\mathbf{C}\right) \quad \text{s.t.} \quad \mathbf{C}^\top\mathbf{C} = \mathbf{I} \quad \text{and} \quad \mathbf{C} \text{ as in (5.4)} \tag{5.6}$$

To understand why this equivalence holds, one should simply note that

$$\text{tr}\left(\mathbf{C}^\top\mathbf{L}\mathbf{C}\right) = \sum_{\ell=1}^{k}\frac{1}{|\mathsf{V}_\ell|}\mathbf{z}_\ell^\top\mathbf{L}\mathbf{z}_\ell = \sum_{\ell=1}^{k}\frac{1}{|\mathsf{V}_\ell|}\sum_{i>j}\mathbf{W}(i, j)(\mathbf{z}_\ell(i) - \mathbf{z}_\ell(j))^2$$
$$= \sum_{\ell=1}^{k}\frac{w(\mathsf{V}_\ell, \bar{\mathsf{V}}_\ell)}{|\mathsf{V}_\ell|} = 2\,\texttt{rcut}(\mathscr{P}).$$

Solving (5.6) is obviously still NP-hard as the only thing we have achieved is to rewrite the rcut minimization problem in matrix form. Yet, in this form, it is easier to realize that one may find an approximate solution by relaxing the discreteness constraint "**C** as in (5.4)." In the absence of the hard-to-deal-with constraint, the relaxed problem is not only polynomially solvable but also possesses a closed-form solution! By the Courant–Fischer–Weyl (min-max) theorem, the solution is given by the first k eigenvectors $\mathbf{U}_k = [\mathbf{u}_1, \mathbf{u}_2, \ldots, \mathbf{u}_k]$ of $\mathbf{L}$:

$$\mathbf{U}_k = \arg\min_{\mathbf{C} \in \mathbb{R}^{n \times k}} \operatorname{tr}\left(\mathbf{C}^\top \mathbf{L} \mathbf{C}\right) \quad \text{subject to} \quad \mathbf{C}^\top \mathbf{C} = \mathbf{I}.$$

This relaxation is not unique to the combinatorial Laplacian. In the same spirit, the minimum `ncut` optimization problem can be formulated in terms of the normalized Laplacian matrix $\mathbf{L}_n$, and the relaxed problem's solution is given by the first k eigenvectors of $\mathbf{L}_n$.

A difficulty still lies before us: how do we go from a real-valued $\mathbf{U}_k$ to a partition of the nodes? The two next subsections aim to motivate the use of k-means as a rounding heuristic. The exposition starts from the simple case when there are only two clusters ($k = 2$) before considering the general case (arbitrary k).

5.2.2.1 The Case of Two Clusters: Thresholding Suffices

For simplicity, we first consider the case of two clusters. If one constructs a partitioning $\mathscr{P}_t$ with $\mathsf{V}_1 = \{v_i : \mathbf{u}_2(i) > t\}$ and $\mathsf{V}_2 = \{v_i : \mathbf{u}_2(i) \leq t\}$ for every level set $t \in (-1, 1)$, then it is a folklore result that

$$\texttt{rcut}(\mathscr{P}^*) \leq \min_t \texttt{rcut}(\mathscr{P}_t) \leq 2\sqrt{\texttt{rcut}(\mathscr{P}^*)\left(d_{\max} - \frac{\lambda_2}{2}\right)}, \tag{5.7}$$

with $\mathscr{P}^* = \arg\min_{\mathscr{P}} \texttt{rcut}(\mathscr{P})$ being the optimal partitioning, $d_{\max}$ is the maximum degree of any node in V, and λ_2 the second smallest eigenvalue of $\mathbf{L}$. The upper bound is achieved by the tree-cross-path graph constructed by Guattery and Miller [57]. In an analogous manner, if $\mathscr{P}^* = \arg\min_{\mathscr{P}} \texttt{ncut}(\mathscr{P})$ is the optimal partitioning w.r.t. the `ncut` cost and every $\mathscr{P}_t$ has been constructed by thresholding the second eigenvector of $\mathbf{L}_n$, then

$$\texttt{ncut}(\mathscr{P}^*) \leq \min_t \texttt{ncut}(\mathscr{P}_t) \leq 2\sqrt{\texttt{ncut}(\mathscr{P}^*)}. \tag{5.8}$$

Inequality (5.8) can be derived as a consequence of the *Cheeger inequality*, a key result of spectral graph theory [32], which for the normalized Laplacian reads:

$$\frac{\lambda_2}{2} \leq \texttt{ncut}(\mathscr{P}^*) \leq \min_{\mathsf{V}} \frac{w(\mathsf{V}, \bar{\mathsf{V}})}{\min\{w(\mathsf{V}), w(\bar{\mathsf{V}})\}} \leq \min_t \texttt{ncut}(\mathscr{P}_t) \leq \sqrt{2\lambda_2}.$$

As a consequence, we have

$$\texttt{ncut}(\mathscr{P}^*) \le \min_t \texttt{ncut}(\mathscr{P}_t) \le \sqrt{2\lambda_2} \le \sqrt{4\texttt{ncut}(\mathscr{P}^*)} = 2\sqrt{\texttt{ncut}(\mathscr{P}^*)},$$

as desired. The derivation of the rcut bound given in (5.7) follows similarly.

5.2.2.2 More Than Two Clusters: Use *k*-Means

As the number of clusters k increases, the brute-force approach of testing every level set becomes quickly prohibitive. But why is k-means the right way to obtain the clusters in the spectral embedding? Though a plethora of experimental evidence advocate the use of k-means, a rigorous justification is still lacking. The interested reader may refer to [83] for an example of an analysis of spectral partitioning without k-means.

More recently, Peng et al. [107] came up with a mathematical argument showing that, if G is well clusterable and we use a k-means algorithm (e.g., [76]) which guarantees that the identified solution $\tilde{\mathsf{C}}$ abides to

$$f(\tilde{\mathsf{C}}; \mathsf{X}) \le (1+\epsilon) f(\mathsf{C}^*; \mathsf{X}),$$

where C^* is the optimal solution of the k-means problem, then the partitioning $\tilde{\mathscr{P}}$ produced by spectral clustering when using $\mathbf{L}_n$ has ncut cost provably close to that of the optimal partitioning $\mathscr{P}^*$. In particular, it was shown that, as long as $\lambda_{k+1} \ge \mathrm{c}k^2\texttt{ncut}(\mathscr{P}^*)$, then

$$\texttt{ncut}(\mathscr{P}^*) \le \texttt{ncut}(\tilde{\mathscr{P}}) \le \zeta\, \texttt{ncut}(\mathscr{P}^*) \left(1 + \epsilon \frac{k^3}{\lambda_{k+1}}\right),$$

for some constants c, $\zeta > 0$ that are independent of n and k (see also [71]). Note that, using the higher-order Cheeger inequality [83] $\lambda_k/2 \le \texttt{ncut}(\mathscr{P}^*)$, the condition $\lambda_{k+1} \ge \mathrm{c}k^2\texttt{ncut}(\mathscr{P}^*)$ implies

$$\frac{\lambda_{k+1}}{\lambda_k} \ge \frac{\mathrm{c}k^2}{2} = \Omega(k^2).$$

Though hopefully milder than this one,[6] such gap assumptions are very common in the analysis of spectral clustering. Simply put, the larger the gap $\lambda_{k+1} - \lambda_k$ is, the stronger the cluster structure and the easier it is to identify a good clustering.

[6] To construct an example possibly verifying such a strong gap assumption, consider k cliques of size k connected together via only $k-1$ edges, so as to form a loosely connected chain. Even though this is a straightforward clustering problem known to be easy for spectral clustering algorithms, the above theorem's assumption implies $\lambda_{k+1} = \Omega(k^2\texttt{ncut}(\mathscr{P}^*)) = \Omega(k)$ which, independently of n, can only be satisfied when k is a small (recall that the eigenvalues of $\mathbf{L}_n$ are necessarily between 0 and 2).

Besides quantifying the difficulty of the clustering problem, the gap also encodes the robustness of the spectral embedding to errors induced by approximation algorithms [36]. The eigenvectors of a *perturbed* Hermitian matrix exhibit an interesting property: instead of being arbitrary, induced changes are localized w.r.t. the eigenvalue axis, following an inverse square eigenvalue-distance law [89]. More precisely, if $\tilde{\mathbf{u}}_i$ is the i-th eigenvector after perturbation, then the inner products $(\tilde{\mathbf{u}}_i^\top \mathbf{u}_j)^2$ decrease proportionally with $|\lambda_i - \lambda_j|^2$. As such, demanding that $\lambda_{k+1} - \lambda_k$ is large is often helpful in the analysis of spectral clustering algorithms in order to ensure that the majority of useful information (contained within $\mathbf{U}_k$) is preserved (in $\tilde{\mathbf{U}}_k$) despite approximation errors.[7]

5.2.2.3 Choice of Relaxation

The presented relaxation approach is not unique and other relaxations could be equally valid (see, for instance, [17, 24, 112]). This relaxation has nevertheless the double advantage of being theoretically simple and computationally easy to implement. Also, justification of spectral clustering algorithms does not only come from this graph cut perspective and in fact encompasses several approaches that we will not detail here: perturbation approaches or hitting time considerations [138], a polarization theorem [23], consistency derivations [84, 135], etc. Interestingly, recent studies (for instance, [18]) on the stochastic block models have shown that spectral clustering (on other matrices than the Laplacian, such as the non-backtracking matrix [73], or the Bethe–Hessian matrix [114] or other similar deformed Laplacians [34]) perform well up to the detectability threshold of the block structure.

5.2.3 *Computational Complexity Considerations*

What is the computational complexity of spectral clustering as a function of the number of points n, their dimension d, and the number of desired clusters k? Let us examine the three steps involved one by one.

The first step entails the construction of a sparse similarity graph from the input points, which is dominated by the kernel computation and costs $\mathcal{O}(dn^2)$. In the second step, given the graph G consisting of n nodes and e edges.[8] one needs to compute the spectral embedding (step 2 of Algorithm 1). Without exploiting the special structure of a graph Laplacian—other than its sparsity that is—there are two main options:

[7]Usually, one needs to ensure that $\sum_{i \le k, j > k} (\tilde{\mathbf{u}}_i^\top \mathbf{u}_j)^2 / k$ remains bounded.

[8]With e of the order of n if the sparsification step was well conducted.

- Using power iterations, one may identify sequentially each non-trivial eigenvector $\mathbf{u}_\ell$ in time $\mathcal{O}(e/\delta_\ell)$, where $\delta_\ell = \lambda_\ell - \lambda_{\ell-1}$ is the ℓ-th eigenvalue gap and e is the number of edges of the graph [136]. Computing the spectral embedding therefore takes $\mathcal{O}(ke/\delta)$ with $\delta = \min_\ell \delta_\ell$. Unfortunately, there exist graphs[9] such that $\delta = \mathcal{O}(1/n)$, bringing the overall worst-case complexity to $\mathcal{O}(kne)$.
- The Lanczos method can be used to approximate the first k eigenvectors in roughly $\mathcal{O}(ek + nk^2)$ time. This procedure is often numerically unstable resulting to a loss of orthogonality in the computed Krylov subspace basis. The most common way to circumvent this problem is by implicit restart [26], whose computational complexity is not easily derived. The number of restarts, empirically, depends heavily on the eigenvalue distribution in the vicinity of λ_k: if λ_k is in an eigenvalue bulk, the algorithms takes longer than when λ_k is isolated. We decide to write the complexity of restarted Arnoldi as $\mathcal{O}(t(ek+nk^2))$ with t modeling the number of restarts. Note that throughout this paper, t will generically refer to a number of iterations in algorithm complexities. We refer the interested reader to [13] for an in-depth discussion of Lanczos methods.

The third step entails solving the k-means problem, typically by using the Lloyd-Max algorithm to converge to a local minimum of $f(\mathsf{C}; \mathsf{X})$. Since there is no guarantee that this procedure will find a good local minimum, it is usually rerun multiple times, starting in each case from randomly selected centroids C_{ini}. The computational complexity of this third step is $\mathcal{O}(tnk^2)$, where t is a bound on the number of iterations required until convergence multiplied by the number of retries (typically 10).

5.2.4 A Taxonomy of Sampling Methods for Spectral Clustering

For the remainder of the chapter, we propose to classify sampling methods aiming at accelerating one or more of these three steps according to when they sample. If they sample before step 1, they are detailed in Sect. 5.3. Methods that assume that the similarity graph is given or well-approximated and sample between steps 1 and 2 will be found in Sect. 5.4. Finally, methods that assume that the spectral embedding has been exactly computed or well-approximated and sample before the k-means step are explained in Sect. 5.5. This classification of methods, like all classification systems, bears a few flaws. For instance, Nyström methods can be applied to both the context of Sects. 5.3 and 5.4 and are thus mentioned in both. Also, we decided to include the pseudo-code of only a few chosen algorithms that we think are illustrative of the literature. This choice is of course subjective and

[9]The combinatorial Laplacian of a complete balanced binary tree on $k \geq 3$ levels and $n = 2^k - 1$ nodes has $\frac{1}{n} \leq \lambda_2 \leq \frac{2}{n}$ [56].

debatable. Notwithstanding these flaws, we hope that this taxonomy clarifies the landscape of existing methods.

5.3 Sampling in the Original Feature Space

This section is devoted to methods that ambitiously aim to reduce the dimension of the spectral clustering problem even before the graph has been formed. Indeed, the naive way of building the similarity graph (step 1 of spectral clustering algorithms) costs $\mathcal{O}(dn^2)$ and, as such, is one of the main computational bottlenecks of spectral clustering. It should be remarked that the present discussion fits into the wider realm of kernel approximation, a proper review of which cannot fit in this chapter: we will thus concentrate on methods that were in practice used for spectral clustering.

5.3.1 Nyström-Based Methods

The methods of this section aim to obtain an approximation $\tilde{\mathbf{U}}_k$ of the exact spectral embedding $\mathbf{U}_k$ via a sampling procedure in the original feature space.

The Nyström method is a well-known algorithm for obtaining a rough low-rank approximation of a positive semi-definite (PSD) matrix $\mathbf{A}$. Here is a high level description of the steps entailed:

Algorithm 3. Nyström's method
Input. PSD matrix $\mathbf{A} \in \mathbb{R}^{n\times n}$, number of samples m, desired rank k

1. Let S be m column indices chosen by some sampling procedure.
2. Denote by $\mathbf{B} = \mathbf{A}(\mathsf{S}, \mathsf{S}) \in \mathbb{R}^{m\times m}$ and $\mathbf{C} = \mathbf{A}(:, \mathsf{S}) \in \mathbb{R}^{n\times m}$ the sub-matrices indexed by S.
3. Let $\mathbf{B} = \mathbf{Q}\boldsymbol{\Sigma}\mathbf{Q}^\top$ be the eigen-decomposition of $\mathbf{B}$ with the diagonal of $\boldsymbol{\Sigma}$ sorted in decreasing magnitude.
4. Compute the rank-k approximation of $\mathbf{B}$ as $\mathbf{B}_k = \mathbf{Q}_k\boldsymbol{\Sigma}_k\mathbf{Q}_k^\top$, where $\mathbf{Q}_k = \mathbf{Q}(:, :k) \in \mathbb{R}^{n\times k}$ and $\boldsymbol{\Sigma}_k = \boldsymbol{\Sigma}(:k, :k)$.

Possible outputs:

- A low-rank approximation $\tilde{\mathbf{A}} = \mathbf{C}\mathbf{B}^+\mathbf{C}^\top \in \mathbb{R}^{n\times n}$ of $\mathbf{A}$
- A rank-k approximation $\tilde{\mathbf{A}}_k = \mathbf{C}\mathbf{B}_k^+\mathbf{C}^\top \in \mathbb{R}^{n\times n}$ of $\mathbf{A}$
- The top k eigenvectors of $\tilde{\mathbf{A}}_k$, stacked as columns in matrix $\tilde{\mathbf{V}}_k \in \mathbb{R}^{n\times k}$, obtained by orthonormalizing the columns of $\tilde{\mathbf{Q}}_k = \mathbf{C}\mathbf{Q}_k\boldsymbol{\Sigma}_k^{-1} \in \mathbb{R}^{n\times k}$

Various guarantees are known for the quality of $\tilde{\mathbf{A}}$ depending on the type of sampling utilized (i.e., how the indices in S are selected in step 1) and the preferred notion of error (spectral $\|.\|_2$ vs Frobenius $\|.\|_F$ vs trace $\|.\|_*$ norm) [50, 54, 77, 148]. For instance:

Theorem 5.1 (Lemma 8 for $q = 1$ in [54]) *Let $\epsilon \in (0, 1)$ and $\delta \in (0, 1)$ and suppose that S contains the indices of m columns drawn i.i.d. uniformly at random (with or without replacement). Then:*

$$\|\mathbf{A} - \tilde{\mathbf{A}}\|_2 \leq \left(1 + \frac{n}{(1-\epsilon)m}\right) \|\mathbf{A} - \mathbf{A}_k\|_2$$

holds with probability at least $1 - 3\delta$, provided that $m \geq 2\epsilon^{-2}\mu k \log(k/\delta)$, where

$$\mu = \frac{n}{k} \max_{i=1,\ldots,n} \|\mathbf{V}_k(i,:)\|_2^2$$

is the coherence associated with the first k eigenvectors $\mathbf{V}_k$ of $\mathbf{A}$, and $\mathbf{A}_k$ is the best rank-k approximation of $\mathbf{A}$.

Guarantees independent of the coherence can be obtained for more advanced sampling methods. Perhaps the most well-known method is that of *leverage scores*, where one draws m samples independently by selecting (with replacement) the i-th column with probability $\mathbf{p}_i = \|\mathbf{V}_k(i,:)\|_2^2/k$.

Theorem 5.2 (Lemma 5 for $q = 1$ in [54]) *Let $\epsilon \in (0, 1)$ and $\delta \in (0, 1)$ and suppose that S contains the indices of m columns drawn i.i.d. with replacement from such a probability distribution. Then:*

$$\|\mathbf{A} - \tilde{\mathbf{A}}\|_2 \leq \|\mathbf{A} - \mathbf{A}_k\|_2 + \epsilon^2 \|\mathbf{A} - \mathbf{A}_k\|_*$$

holds with probability at least $0.8 - 2\delta$ provided that $m \geq \mathcal{O}(\epsilon^{-2} k \log(k/\delta))$.

Computing leverage scores exactly is computationally prohibitive since it necessitates a partial SVD decomposition of $\mathbf{A}$, which we are trying to avoid in the first place. Nevertheless, it is possible to approximate all leverage scores with a multiplicative error guarantee in time roughly $\mathcal{O}(ek\log(e))$ if $\mathbf{A}$ has $O(e)$ non-zero entries. (see Algorithms 1–3 in [54]). Many variants of the above exist [77, 78], but to the best of our knowledge, the fastest current Nyström algorithm utilizes ridge leverage scores with a complex recursive sampling scheme and runs in time nearly linear in n [100].

Nyström for Spectral Clustering Though initially conceived for low-rank approximation, Nyström's method can also be used to accelerate spectral clustering. The key observation is that $\mathbf{U}_k$, the tailing k eigenvectors of the graph representative matrix $\mathbf{R}$, can be interpreted as the top k eigenvectors of the PSD matrix $\mathbf{A} = \|\mathbf{R}\|_2\mathbf{I} - \mathbf{R}$. As such, the span of the k top eigenvectors of $\tilde{\mathbf{A}}_k$ obtained by running

Algorithm 3 on $\mathbf{A}$ is an approximation of the span of the exact spectral embedding. Different variants of this idea have been considered for the acceleration of spectral clustering [19, 48, 85, 86, 97, 141].

Following our taxonomy, we hereby focus on the case where we have at our disposal n points $\mathbf{p}_i$ in dimension d, and the similarity graph has yet to be formed. The case where the graph is known is deferred to Sect. 5.4.

In this case, we cannot run Algorithm 3 on $\mathbf{A} = \|\mathbf{R}\|_2\mathbf{I} - \mathbf{R}$ as the graph, and *a fortiori* its representative matrix $\mathbf{R}$ has not yet been formed. What we *can* have access to *efficiently* is $\mathbf{B} = s(\mathbf{K}(\mathsf{S}, \mathsf{S}))$ and $\mathbf{C} = s(\mathbf{K}(:, \mathsf{S}))$, as these require only a partial computation of the kernel and cost only $\mathcal{O}(dnm)$. Note that s is a sparsification function that is applied on a subset of the kernel matrix.

The following pseudo-code exemplifies how Nyström-based techniques can be used to approximate the first k eigenvectors $\mathbf{U}_k$ associated with the normalized Laplacian matrix (i.e., here $\mathbf{R} = \mathbf{L}_n$):

Algorithm 3b. Nyström for spectral clustering [85]
Input. The set of points P, the number of desired clusters k, a sampling set S of size $m \geq k$

1. Compute the sub-matrices $\mathbf{B}{=}s(\mathbf{K}(\mathsf{S}, \mathsf{S})) \in \mathbb{R}^{m\times m}$ and $\mathbf{C} = s(\mathbf{K}(:, \mathsf{S})) \in \mathbb{R}^{n\times m}$, where s is a sparsification function.
2. Let $\mathbf{D}_r = \text{diag}(\mathbf{B1})$ be the $m \times m$ degree matrix.
3. Compute the top k eigenvalues $\boldsymbol{\Sigma}_k$ and eigenvectors $\mathbf{Q}_k$ of $\mathbf{D}_r^{-1/2}\mathbf{B}\mathbf{D}_r^{-1/2}$.
4. Set $\tilde{\mathbf{Q}}_k = \mathbf{C}\mathbf{D}_r^{-1/2}\mathbf{Q}_k\boldsymbol{\Sigma}_k^{-1}$.
5. Let $\mathbf{D}_l = \text{diag}(\tilde{\mathbf{Q}}_k\boldsymbol{\Sigma}_k\tilde{\mathbf{Q}}_k^\top\mathbf{1})$ be the $n \times n$ degree matrix.
6. Compute $\tilde{\mathbf{U}}_k$ obtained by orthogonalizing $\mathbf{D}_l^{-1/2}\tilde{\mathbf{Q}}_k$.

Output: $\tilde{\mathbf{U}}_k$, an approximation of the spectral embedding $\mathbf{U}_k$.

This algorithm runs in $\mathcal{O}(nm\max(d, k))$ time, which is small when m depends mildly on the other parameters of interest. Nevertheless, the algorithm (and others like it) suffers from several issues:

- Algorithm 3b attempts to use Nyström's method on $\mathbf{A} = 2\mathbf{I} - \mathbf{L}_n = \mathbf{I} + \mathbf{D}^{-\frac{1}{2}}s(\mathbf{K})\mathbf{D}^{-\frac{1}{2}}$ via the exact computation of two sub-matrices of $\mathbf{K}$. In doing so, it makes two strong (and uncontrolled) approximations. First of all, the sparsification step (step 1 in Algorithm 3b) is applied to the sub-matrices $\mathbf{K}(\mathsf{S}, \mathsf{S})$ and $\mathbf{K}(:, \mathsf{S})$, deviating from the correct sparsification procedure that takes into account the entire kernel matrix $\mathbf{K}$. Second, the degree matrix $\mathbf{D}$ is never exactly computed as knowing it exactly would entail computing exactly $s(\mathbf{K})$, which is precisely what we are trying to avoid. Existing methods thus rely on heuristic approximations of the degree in order to bypass this difficulty (see steps 2 and 5 of Algorithm 3b).

- Since we do not have direct access to the kernel matrix, we cannot utilize advanced sampling methods such as leverage scores to draw the sampling set S. This is particularly problematic if (due to sparsification) matrices **B** and **C** are sparse, as for sparse matrices uniform sampling is known to perform poorly [97]. Techniques that rely on distances between columns do not fair much better. Landmark-based approaches commonly perform better in simple problems but suffer when the clusters are non-convex [19]. We refer the reader to the work by Mohan et al. [97] for more information on landmark-based methods. The latter work also describes an involved sampling scheme that is aimed at general (i.e., non-convex) clusters.

For the reasons highlighted above, the low-rank approximation guarantees accompanying the classical Nyström method cannot be directly used here. *A fortiori*, it is an open question how much the quality of the spectral clustering solution is affected by using the centroids obtained by running k-means on $\tilde{\mathbf{U}}_k$.

Column Sampling Akin in spirit to Nyström methods, an alternative approach to accelerating spectral clustering was inspired by column sampling low-rank approximation techniques [37, 43].

An instance of such algorithms was put forth under the name of cSPEC (column sampling spectral clustering) by Wang et al. [141]. Let $\mathbf{C} = \mathbf{U}_C \mathbf{\Sigma}_C \mathbf{V}_C^\top$ be the singular value decomposition of the $n \times m$ matrix $\mathbf{C} = s(\mathbf{K}(:, \mathsf{S}))$. Then, matrices

$$\tilde{\mathbf{\Sigma}} = \sqrt{\frac{n}{m}}\, \mathbf{\Sigma}_C \quad \text{and} \quad \tilde{\mathbf{U}} = \mathbf{C}\mathbf{V}_C \mathbf{\Sigma}_C^+$$

are interpreted as an approximation of the actual eigenvalues and eigenvectors of **K** and thus $\mathbf{U}_k$ can be substituted by the first k columns of $\tilde{\mathbf{U}}$. This algorithm runs in $\mathcal{O}(ndm + nm^2)$.

Authors in [29] propose a hybrid method, between column sampling and the representative-based methods discussed in Sect. 5.3.3, where they propose the following approximate factorization of the data matrix:

$$(\mathbf{p}_1 | \dots | \mathbf{p}_n) \simeq \mathbf{F}\mathbf{Z} \in \mathbb{R}^{d \times n}, \tag{5.9}$$

where $\mathbf{F} \in \mathbb{R}^{d \times m}$ concatenates the feature vectors of m sampled points and $\mathbf{Z} \in \mathbb{R}^{m \times n}$ represents all unsampled points as approximate linear combinations of the representatives, computed via sparse coding techniques [82].[10] The SVD of $\tilde{\mathbf{D}}^{-1/2}\mathbf{Z}$, with $\tilde{\mathbf{D}}$ the row-sum of **Z**, is then computed to obtain an approximation $\tilde{\mathbf{U}}_k$ of $\mathbf{U}_k$. The complexity of their algorithm is also $\mathcal{O}(ndm + nm^2)$.

[10] Authors in [116] have a very similar proposition as [29], adding a projection phase at the beginning to reduce the dimension d (see Sect. 5.3.4.2). Similar ideas may also be found in [137].

In these methods, the choice of the sample set S is, of course, central and has been much debated. Popular options are uniformly at random or via better-tailored probability distributions, via a first k-means (with $k = m$) pass on P, or via other selective sampling methods. Also, as with most extensions of Nyström's method to spectral clustering, column sampling methods for spectral clustering do not come with end-to-end approximation guarantees on $\mathbf{U}_k$.

In the world of low-rank matrix approximation the situation is somewhat more advanced. Recent work in column sampling utilizes adaptive sampling with leverage scores in time $\mathcal{O}(e + n\text{poly}(k))$, or uniformly i.i.d. after preconditioning by a fast randomized Hadamard transform [41, 145]. Others have also used a correlated version called volume sampling to obtain column indices [37]. Nevertheless, this literature extends beyond the scope of this chapter and thus we invite the interested reader to consider the aforementioned references for a more in-depth perspective.

5.3.2 Random Fourier Features

Out of several sketching techniques one could *a priori* use to accelerate spectral clustering, we focus on random Fourier features (RFF) [110]: a method that samples in the Fourier space associated with the original feature space. Even though RFFs have originally been developed to approximate a kernel matrix $\mathbf{K}$ in time linear in n instead of the quadratic time necessary for its exact computation, they can in fact be used to obtain an approximation $\tilde{\mathbf{U}}_k$ of the exact spectral embedding $\mathbf{U}_k$.

Let us denote by κ the RBF kernel, i.e., $\kappa(\mathbf{t}) = \exp(-\mathbf{t}^2/\sigma^2)$, whose Fourier transform is:

$$\hat{\kappa}(\boldsymbol{\omega}) = \int_{\mathbb{R}^d} \kappa(\mathbf{t}) \exp^{-i\boldsymbol{\omega}^\top \mathbf{t}} \mathrm{d}\mathbf{t}. \tag{5.10}$$

The above takes real values as κ is symmetric. One may write:

$$\kappa(\mathbf{p}, \mathbf{q}) = \kappa(\mathbf{p} - \mathbf{q}) = \frac{1}{Z} \int_{\mathbb{R}^d} \hat{\kappa}(\boldsymbol{\omega}) \exp^{i\boldsymbol{\omega}^\top (\mathbf{p}-\mathbf{q})} \mathrm{d}\boldsymbol{\omega}, \tag{5.11}$$

where, in order to ensure that $\kappa(\mathbf{p}, \mathbf{p}) = 1$, the normalization constant is set to $Z = \int_{\mathbb{R}^d} \hat{\kappa}(\boldsymbol{\omega})\mathrm{d}\boldsymbol{\omega}$. According to Bochner's theorem, and due to the fact that κ is positive-definite, $\hat{\kappa}/Z$ is a valid probability density function. $\kappa(\mathbf{p}, \mathbf{q})$ may thus be interpreted as the expected value of $\exp^{i\boldsymbol{\omega}^\top (\mathbf{p}-\mathbf{q})}$ provided that $\boldsymbol{\omega}$ is drawn from $\hat{\kappa}/Z$:

$$\kappa(\mathbf{p}, \mathbf{q}) = \mathbb{E}_{\boldsymbol{\omega}} \left(\exp^{i\boldsymbol{\omega}^\top (\mathbf{p}-\mathbf{q})} \right) \tag{5.12}$$

Drawing ω from the distribution $\hat{\kappa}/Z$ is equivalent to drawing independently each of its d entries according to the normal law of mean 0 and variance $2/\sigma^2$. Indeed: $\hat{\kappa}(\boldsymbol{\omega}) = \pi^{d/2}\sigma^d \exp(-\sigma^2\boldsymbol{\omega}^2/4)$ and $Z = \int_{\mathbb{R}^d} \hat{\kappa}(\boldsymbol{\omega})\mathrm{d}\boldsymbol{\omega} = (2\pi)^d$, leading to

$$\frac{\hat{\kappa}(\boldsymbol{\omega})}{Z} = \left(\frac{\sigma}{2\sqrt{\pi}}\right)^d \exp^{-\sigma^2\omega^2/4}.$$

In practice, we draw independently m such vectors ω to obtain the set of sampled frequencies $\Omega = (\boldsymbol{\omega}_1, \ldots, \boldsymbol{\omega}_m)$. For each data point $\mathbf{p}_i$, and given this set of samples Ω, we define the associated random Fourier feature vector:

$$\boldsymbol{\psi}_i = \frac{1}{\sqrt{m}}[\cos(\boldsymbol{\omega}_1^\top \mathbf{p}_i)| \cdots | \cos(\boldsymbol{\omega}_m^\top \mathbf{p}_i)| \sin(\boldsymbol{\omega}_1^\top \mathbf{p}_i)| \cdots | \sin(\boldsymbol{\omega}_m^\top \mathbf{p}_i)]^\top \in \mathbb{R}^{2m}, \tag{5.13}$$

and call $\boldsymbol{\Psi} = (\boldsymbol{\psi}_1 | \cdots | \boldsymbol{\psi}_n) \in \mathbb{R}^{2m \times n}$ the RFF matrix. Other embeddings are possible in the RFF framework, but this one was shown to be the most appropriate to the Gaussian kernel [127]. As m increases, $\boldsymbol{\psi}_i^\top \boldsymbol{\psi}_j$ concentrates around its expected value $\kappa(\mathbf{p}_i, \mathbf{p}_j)$: $\boldsymbol{\psi}_i^\top \boldsymbol{\psi}_j \simeq \kappa(\mathbf{p}_i, \mathbf{p}_j)$. Proposition 1 of [127] states the tightness of this concentration: it shows that the approximation starts to be valid with high probability for $m \geq \mathcal{O}(d \log d)$. The Gaussian kernel matrix is thus well approximated as $\mathbf{K} \simeq \boldsymbol{\Psi}^\top \boldsymbol{\Psi}$. With such a low-rank approximation $\boldsymbol{\Psi}$ of $\mathbf{K}$, one can estimate the degrees,[11] degree-normalize $\boldsymbol{\Psi}$ to obtain a low-rank approximation of the normalized Laplacian $\mathbf{L}_n$, and perform an SVD to directly obtain an approximation $\tilde{\mathbf{U}}_k$ of the spectral embedding $\mathbf{U}_k$. The total cost to obtain this approximation is $\mathcal{O}(ndm+nm^2)$. These ideas were developed in Refs. [31, 146], for instance.

As in Nyström methods however, the concentration guarantees of RFFs for $\mathbf{K}$ do not extend to the degree-normalized case; moreover, the sparsification step 1b of spectral clustering is ignored. Note that improving over RFFs in terms of efficiency and concentration properties is the subject of recent research (see, for instance, [81]).

5.3.3 The Paradigm of Representative Points

The methods detailed here sample in the original feature space and directly obtain a control on the misclustering rate due to the sampling process. They are based on the following framework:

1. Sample m so-called representatives.
2. Run spectral clustering on the representatives.
3. Lift the solution back to the entire dataset.

---

[11] An approximation of the degree d_i of node v_i is $\psi_i^\top \bar{\psi}$ where $\bar{\psi} = \sum_j \psi_j$. All degrees can thus be estimated in time $\mathcal{O}(nm^2)$.

Let us illustrate this with the example of KASP:

Algorithm 4. KASP: k-means-based approximate spectral clustering [147]
Input. A set of n points $\mathsf{P} = (\mathbf{p}_1, \mathbf{p}_2, \ldots, \mathbf{p}_n)$ in dimension d, a number of desired clusters k, and a number of representatives m.

1. Perform k-means with $k = m$ on P and obtain:
   (a) the cluster centroids $\mathsf{Y} = (\mathbf{y}_1, \ldots, \mathbf{y}_m)$ as the m representative points.
   (b) a correspondence table to associate each $\mathbf{p}_i$ to its nearest representative
2. Run spectral clustering on Y to get the cluster membership of each $\mathbf{y}_i$.
3. Lift the cluster membership to each $\mathbf{p}_i$ by looking up the cluster membership of its representative in the correspondence table.

Output: k clusters

The complexity of KASP is bounded by[12] $\mathcal{O}(mdnt + m^3)$. For a summary of the analysis given in [147], let us consider the cluster memberships given by exact spectral clustering on P as well as the memberships given by exact spectral clustering on $\tilde{\mathsf{P}} = (\mathbf{p}_1 + \epsilon_1, \ldots, \mathbf{p}_n + \epsilon_n)$, where the ϵ_i are any small perturbations on the initial points. Let us denote by $\mathbf{L}_n$ (resp. $\tilde{\mathbf{L}}_n$) the normalized Laplacian matrix of the similarity graph on P (resp. $\tilde{\mathsf{P}}$). The analysis concentrates on the study of the miss-clustering rate ρ:

$$\rho = \frac{\text{\# of points with different memberships}}{n}. \tag{5.14}$$

The main result, building upon preliminary work in [63], stems from a perturbation approach and reads:

Theorem 5.3 *Under the assumptions of Theorem 3 in [147]: $\rho \leq \mathcal{O}\left(\frac{k}{g_0^2}\|\mathbf{L}_n - \tilde{\mathbf{L}}_n\|_F\right)$, where g_0 is a value depending on the spectral gap. Also, under the assumptions of Theorem 6 in [147], one has, with high probability:*

$$\|\mathbf{L}_n - \tilde{\mathbf{L}}_n\|_F \leq \mathcal{O}\left(\sigma_\epsilon^{(2)} + \sigma_\epsilon^{(4)}\right), \tag{5.15}$$

with $\sigma_\epsilon^{(2)}$ and $\sigma_\epsilon^{(4)}$ the 2nd and 4th moments of the perturbation's norms $\|\epsilon_i\|_2$.

[12]It is in fact $\mathcal{O}(mdnt)$ for step 1, and bounded by $\mathcal{O}(dm^2 + m^2k + mk^2)$ for step 2. As $n \geq m$ and $m \geq k$, the total complexity is bounded by $\mathcal{O}(mdnt + m^3)$.

Combining both bounds, one obtains an upper bound on the misclustering rate that depends on the second and fourth moments of the perturbation's norms $\|\epsilon_i\|_2$. The "collapse" of points onto the m representative points, interpreted as a perturbation on the original points, should thus tend to minimize these two moments, leading the authors to propose *distortion-minimizing* algorithms, such as KASP. A very similar algorithm, eSPEC, is described in [141].

5.3.4 Other Methods

5.3.4.1 Approximate Nearest Neighbor Search Algorithms

The objective here is to approximate the nearest neighbor graph efficiently. Even though these methods are not necessarily based on sampling, we include them in the discussion as they are frequently used in practice.

Given the feature vectors $\mathbf{p}_1, \ldots, \mathbf{p}_n \in \mathbb{R}^d$ and a query point $\mathbf{q} \in \mathbb{R}^d$, the exact nearest neighbor search (exact NNS) associated with P and $\mathbf{q}$ is $\mathbf{p}^* = \operatorname{argmin}_{p \in \mathsf{P}} \operatorname{dist}(\mathbf{q}, \mathbf{p})$, where *dist* stands for any distance. Different distances are possible depending on the choice of kernel κ. We will here consider the Euclidean norm as it enters the definition of the classical RBF kernel. Computing the exact NNS costs $\mathcal{O}(dn)$. The goal of the approximate NNS field of research is to provide faster algorithms that have the following control on the error.

Definition 5.1 Point $\mathbf{p}^*$ is an ϵ-approximate nearest neighbor of query $\mathbf{q} \in \mathbb{R}^d$, if

$$\forall \mathbf{p} \in \mathsf{P} \qquad \operatorname{dist}(\mathbf{q}, \mathbf{p}^*) \leq (1+\epsilon) \operatorname{dist}(\mathbf{q}, \mathbf{p}).$$

For $\epsilon = 0$, this reduces to exact NNS.

Extensions of this objective to the k-nearest neighbor goal are considered in the NNS literature. A k-nearest neighbor graph can then be constructed simply by running an approximate k-NNS query for each object $\mathbf{p}_i$. Thus, approximate NSS algorithms are interesting candidates to approximate the adjacency matrix of the nearest-neighbor affinity graph, that we need in step 1 of spectral clustering. Many algorithms exist, their respective performances depending essentially on the dimension d of the feature vectors. According to [9], randomized k-d forests as implemented in the library FLANN [98] are considered state of the art for dimension of around 100, whereas methods based on balanced box decomposition (BBD) [4, 7] are known to perform well for d roughly smaller than 100. In high dimensions, to avoid the curse of dimensionality, successful approaches are, for instance, based on hashing methods (such as locality sensitive hashing (LSH) [5], product quantization (PQ) [66]), or k-d generalized random forests [9]. Finally, proximity graph methods that sequentially improve over a first coarse approximation of the k-NN graph (or other graph structures such as navigable graphs) have received a large attention recently and are becoming state of the art in regimes where quality of approximation primes (see, for instance, [8, 40, 51, 94]). Such tools come with various levels of guarantees and computation costs, the details of which are not in the scope of this chapter.

Experimentally, to obtain an approximate k-NN graph with a typical recall rate[13] of 0.9, these algorithms are observed to achieve a complexity of $\mathcal{O}(dn^\alpha)$ with α close to 1 ($\alpha \simeq 1.1$ in [40], for instance).

5.3.4.2 Feature Selection and Feature Projection

Some methods work on reducing the factor d of the complexity $\mathcal{O}(dn^2)$ of the kernel computation via feature selection, i.e., the sampling of features deemed more useful for the underlying clustering task, or feature projection, i.e., the projection on usually random subspaces of dimension $d' < d$. Feature selection methods are usually designed to *improve* the classification by removing features that are too noisy or useless for the classification. We thus do not detail further these methods as they are not approximation algorithms *per se*. The interested reader will find some entries in the literature via references [25, 35, 60, 149]. Projection methods use random projections of the original points P on spaces of dimension $d' \sim \log n$ in order to take advantage of the Johnson–Lindenstrauss lemma of norm conservation: the kernel computed from the projected features in dimension d' is thus an approximation of the true kernel with high probability. We refer to the works [64, 116] for more details.

5.4 Sampling Given the Similarity Graph

We now suppose that the similarity graph is either given (e.g., in cases where the original data *is* a graph) or has been well approximated (by approximate k-NN search, for instance) and concentrate on sampling-based methods that aim to reduce the cost of computing the first k eigenvectors of $\mathbf{R}$.

These methods predominantly aim to approximate $\mathbf{R}$ by a smaller matrix $\tilde{\mathbf{R}}$ of size m. The eigen-decomposition is done in $\mathbb{R}^m$ which can be significantly cheaper when $m \ll n$. In addition, each method comes with a fast way of lifting vectors from $\mathbb{R}^m$ back to $\mathbb{R}^n$ (this is usually a linear transformation). After lifting, the eigenvectors of $\tilde{\mathbf{R}}$ are used as a proxy for those of $\mathbf{R}$.

Unlike the previous section where a strong approximation guarantee of the exact embedding $\mathbf{U}_k$ by an efficiently computed $\tilde{\mathbf{U}}_k$ was a distant and difficult goal to achieve in itself, we will see in this section that the knowledge of the similarity

[13]The recall rate for a node is the number of correctly identified k-NN divided by k. The recall rate for a k-NN graph is the average recall rate over all nodes.

graph not only enables to obtain such strong approximation guarantees, but also enables to control how the error on $\mathbf{U}_k$ transfers as an error on the k-means cost.

To be more precise, recall (5.1) defining the k-means cost $f(\mathsf{C};\mathsf{X})$ associated with the n points $\mathsf{X} = (\mathbf{x}_1, \ldots, \mathbf{x}_n)$ and a centroid set C. Now, suppose that we have identified a set of n points $\tilde{\mathsf{X}} = (\tilde{\mathbf{x}}_1|\ldots|\tilde{\mathbf{x}}_n)$ that are meant to approximate the exact spectral embedding X. Moreover, let C^* (resp. $\tilde{\mathsf{C}}^*$) be the optimal set of k centroids minimizing the k-means cost on X (resp. $\tilde{\mathsf{X}}$). We will see that several (not all) approximation methods of this section achieve an end-to-end approximation guarantee of the form

$$\left| f(\mathsf{C}^*;\mathsf{X})^{1/2} - f(\tilde{\mathsf{C}}^*;\mathsf{X})^{1/2} \right| \leq \epsilon,$$

for some small ϵ with—at least—constant probability. Such an end-to-end guarantee is indeed more desirable than a simple guarantee on the distance between $\mathbf{U}_k$ and $\tilde{\mathbf{U}}_k$: it informs us on the approximation quality of the attained clustering.

5.4.1 Nyström-Based Methods

The Nyström-based methods discussed in Sect. 5.3.1 are also applicable here. Let us concentrate on the choice $\mathbf{R} = \mathbf{L}_n$ to illustrate the main ideas. As explained in Sect. 5.3.1, the tailing k eigenvectors $\mathbf{U}_k$ of $\mathbf{L}_n$ can be interpreted as the top k eigenvectors of the PSD matrix $\mathbf{A} = 2\mathbf{I} - \mathbf{L}_n$. As such, the span of the top-k eigenvectors of $\tilde{\mathbf{A}}_k$, span$(\tilde{\mathbf{U}}_k)$, obtained by running Algorithm 3 on $\mathbf{A}$ should approximate the span of $\mathbf{U}_k$. Now, how does one go from Nyström theorems such as Theorem 5.2 to error bounds on the k-means cost function?

The first step towards an end-to-end guarantee relies on the following result:

Lemma 1 (See the Proof of Theorem 6 in [21]) *Denote by $\tilde{\mathsf{C}}^*$ the optimal centroid set obtained by solving k-means on the rows of $\tilde{\mathbf{U}}_k$. It holds that*

$$\left| f(\mathsf{C}^*;\mathsf{X})^{1/2} - f(\tilde{\mathsf{C}}^*;\mathsf{X})^{1/2} \right| \leq 2\|\mathbf{E}\|_F, \tag{5.16}$$

where $\mathbf{E} = \mathbf{U}_k\mathbf{U}_k^\top - \tilde{\mathbf{U}}_k\tilde{\mathbf{U}}_k^\top$.

This means that the error made by considering the optimal k-means solution based on $\tilde{\mathbf{U}}_k$ (instead of $\mathbf{U}_k$) is controlled by the Frobenius norm of the projector difference $\mathbf{E} = \mathbf{U}_k\mathbf{U}_k^\top - \tilde{\mathbf{U}}_k\tilde{\mathbf{U}}_k^\top$. Furthermore, since[14] $\|\mathbf{E}\|_F \leq \sqrt{2k}\|\mathbf{E}\|_2$ and

[14]Based on three arguments: (i) for any two matrices $\mathbf{M}_1$ and $\mathbf{M}_2$ of rank r_1 and r_2 it holds that rank$(\mathbf{M}_1 + \mathbf{M}_2) \leq r_1 + r_2$, (ii) for any matrix $\mathbf{M}$ or rank r, $\|\mathbf{M}\|_F \leq \sqrt{r}\|\mathbf{M}\|_2$, and (iii) both $\mathbf{U}_k$ and $\tilde{\mathbf{U}}_k$ are of rank k.

$\|\mathbf{E}\|_2 = \|\sin(\Theta(\mathbf{U}_k, \tilde{\mathbf{U}}_k))\|_2$, we can apply the Davis–Kahan $\sin\Theta$ perturbation theorem (see, for instance, Section VII of [16]) and, provided that $\sigma_k - \tilde{\sigma}_{k+1} > 0$, obtain:

$$\|\mathbf{E}\|_F \leq \sqrt{2k}\|\mathbf{E}\|_2 \leq \sqrt{2k}\,\frac{\|\mathbf{A} - \tilde{\mathbf{A}}\|_2}{\sigma_k - \tilde{\sigma}_{k+1}},$$

where $\{\sigma_i\}$ (resp. $\{\tilde{\sigma}_i\}$) are the singular values of $\mathbf{A}$ (resp. $\tilde{\mathbf{A}}$) ordered decreasingly.[15] The final bound is obtained by combining the above with the leverage score sampling bound given by Theorem 5.2.

Theorem 5.4 *Let $\tilde{\mathbf{U}}_k$ be the eigenvectors obtained by running Algorithm 3 on $\mathbf{A} = 2\mathbf{I} - \mathbf{L}_n$ (with the leverage score sampling scheme for the m samples $\mathcal{S}$ of step 1). Denote by $\tilde{\mathsf{C}}^*$ the optimal centroid set obtained by solving k-means on the rows of $\tilde{\mathbf{U}}_k$. Then, for some constant $C > 1$, we have*

$$\left|f(\mathsf{C}^*;\mathsf{X})^{1/2} - f(\tilde{\mathsf{C}}^*;\mathsf{X})^{1/2}\right| \leq 2\frac{\sqrt{2k}}{\sigma_k - \tilde{\sigma}_{k+1}}\left(\sigma_{k+1}(\mathbf{A}) + \frac{Ck\log(k/\delta)}{m}\sum_{j=k+1}^{n}\sigma_j\right)$$

with probability at least $0.8 - 2\delta$.

Examining the above bound one notices that $2\sqrt{2k}\frac{\sigma_{k+1}(\mathbf{A})}{\sigma_k - \tilde{\sigma}_{k+1}}$ is independent of the number of samples. The incompressibility of this error term emanates from $\mathbf{A}$ being (in general) different from its best low-rank approximation. On the other hand, all remaining error terms can be made independent of k and n by setting

$$m = \mathcal{O}\left(k\sqrt{k}\log k\sum_{j=k+1}^{n}\frac{\sigma_j}{\sigma_k - \tilde{\sigma}_{k+1}}\right).$$

This end-to-end guarantee is not satisfactory for several reasons. First of all, it relies on the assumption $\sigma_k > \tilde{\sigma}_{k+1}$, which is not necessarily true. Moreover, the Davis–Kahan theorem could in theory guarantee $\|\mathbf{E}\|_2 \leq \|\mathbf{A}_k - \tilde{\mathbf{A}}_k\|_2/\sigma_k$ and $\|\mathbf{E}\|_2 \leq \|\mathbf{A} - \tilde{\mathbf{A}}_k\|_2/\sigma_k$, which are stronger than the bound depending on $\|\mathbf{A} - \tilde{\mathbf{A}}\|_2$ that we used. Unfortunately, Nyström approximation theorems do not give controls on $\|\mathbf{A}_k - \tilde{\mathbf{A}}_k\|_2$ nor on $\|\mathbf{A} - \tilde{\mathbf{A}}_k\|_2$, impeding tighter end-to-end bounds.

5.4.2 Graph Coarsening

Inspired by the algebraic multi-grid, researchers realized early on that a natural way to accelerate spectral clustering is by graph coarsening [38, 61, 69]. Here,

[15] Note that, in our setting, $\mathbf{A} = 2\mathbf{I} - \mathbf{L}_n$ and $\sigma_k = 2 - \lambda_k$.

instead of solving the clustering problem directly on G, one may first reduce it to a coarser graph G_c consisting of $m \ll n$ nodes using a multi-level graph coarsening procedure. The expensive eigen-decomposition computation is done at a lower cost on the representative matrix of the small graph and the final spectral embedding is obtained by inexpensively lifting and refining the result.

In the notation of [91], coarsening involves a sequence of $c + 1$ graphs

$$G = G_0 = (V_0, E_0, \mathbf{W}_0) \quad G_1 = (V_1, E_1, \mathbf{W}_1) \quad \cdots \quad G_c = (V_c, E_c, \mathbf{W}_c) \tag{5.17}$$

of decreasing size $n = n_0 > n_1 > \cdots > n_c = m$, where each vertex of G_ℓ represents one of more vertices of $G_{\ell-1}$. To express coarsening in algebraic form, we suppose that $\mathbf{L}(G_0) = \mathbf{L}$ is the combinatorial Laplacian associated with G. We then obtain $\mathbf{L}(G_c)$ by applying the following repeatedly:

$$\mathbf{L}(G_\ell) = \mathbf{P}_\ell^{\mp}\mathbf{L}(G_{\ell-1})\mathbf{P}_\ell^{+}, \tag{5.18}$$

where $\mathbf{P}_\ell \in \mathbb{R}^{n_\ell \times n_{\ell-1}}$ is a matrix with more columns than rows, $\ell = 1, 2, \ldots, c$ is the level of the reduction, and symbol $\mp$ denotes the transposed pseudo-inverse. An eigenvector $\tilde{\mathbf{u}} \in \mathbb{R}^m$ of $\mathbf{L}(G_c)$ is lifted back to $\mathbb{R}^n$ by backwards recursion

$$\tilde{\mathbf{u}}_{\ell-1} = \mathbf{P}_\ell \tilde{\mathbf{u}}_\ell,$$

where $\tilde{\mathbf{u}}_c = \tilde{\mathbf{u}}$.

Matrices $\mathbf{P}_1, \mathbf{P}_2, \ldots, \mathbf{P}_c$ are determined by the transformation performed at each level. Specifically, one should define for each level a surjective map $\varphi_\ell : V_{\ell-1} \to V_\ell$ between the original vertex set $V_{\ell-1}$ and the smaller vertex set V_ℓ. We refer to the set of vertices $V_{\ell-1}^{(r)} \subseteq V_{\ell-1}$ mapped onto the same vertex v_r' of V_ℓ as a *contraction set*:

$$V_{\ell-1}^{(r)} = \{v \in V_{\ell-1} : \varphi_\ell(v) = v_r'\}$$

It is easy to deduce from the above that contraction sets induce a partitioning of $V_{\ell-1}$ into n_ℓ subgraphs, each corresponding to a single vertex of V_ℓ.

Then, for any $v_r' \in V_\ell$ and $v_i \in V_{\ell-1}$, matrices $\mathbf{P}_\ell \in \mathbb{R}^{n_\ell \times n_{\ell-1}}$ and $\mathbf{P}_\ell^+ \in \mathbb{R}^{n_{\ell-1} \times n_\ell}$ are given by:

$$\mathbf{P}_\ell(r, i) = \begin{cases} \frac{1}{|V_{\ell-1}^{(r)}|} & \text{if } v_i \in V_{\ell-1}^{(r)} \\ 0 & \text{otherwise} \end{cases} \quad \text{and} \quad \mathbf{P}_\ell^+(i, r) = \begin{cases} 1 & \text{if } v_i \in V_{\ell-1}^{(r)} \\ 0 & \text{otherwise.} \end{cases}$$

The preceding construction is the only one that guarantees that every $\mathbf{L}(G_\ell)$ will be the combinatorial Laplacian associated with G_ℓ [90].

Note that from a computational perspective the reduction is very efficient and can be carried out in linear time: each coarsening level entails multiplication by a sparse matrix, meaning that $\mathcal{O}(e)$ and $\mathcal{O}(n)$ operations suffice, respectively, to coarsen $\mathbf{L}$ and lift any vector (such as the eigenvectors of $\mathbf{L}(G_c)$) from $\mathbb{R}^m$ back to $\mathbb{R}^n$.

5.4.2.1 Coarsening for Spectral Clustering

Using coarsening effectively boils down to determining for each ℓ how to partition $G_{\ell-1}$ into n_ℓ contraction sets $V_\ell^{(1)}, \ldots, V_\ell^{(n_\ell)}$, such that, after lifting, the first k eigenvectors $\tilde{\mathbf{U}}_k$ of $\mathbf{L}(G_c)$ approximate the spectral embedding $\mathbf{U}_k$ derived from $\mathbf{L}$. Alternatively, one may also solve the k-means problem in the small dimension and only lift the resulting cluster assignments [38]. This scheme is computationally superior but we will not discuss it here as it does not come with any guarantees.

Perhaps the most *simple* (and common) method of forming contraction sets is by the *heavy-edge matching heuristic*—originally developed in the multi-grid literature and first considered for graph partitioning in [69]. This method is derived based on the intuition that the larger the weight of an edge, the less likely it will be that the vertices it connects will reside in different clusters. We should therefore aim to contract pairs of vertices connected by a heavy edge (i.e., of large weight) first. Let us consider this case further. By focusing on edges, we basically constrain ourselves by enforcing that every contraction set $V_\ell^{(r)}$ contains either two nodes connected by an edge or a single node, signifying that said node is chosen to remain as is in the coarser graph. As such, we can reformulate the problem of selecting contraction sets at each level as that of selecting the largest number of edges (to attain the largest reduction), while also striving to make the cumulative sum of selected edge weights as large as possible (giving preference to heavy edges). This is exactly the *maximum weight matching problem*, which can be approximated in linear time [44].

A plethora of numerical evidence motivates the use of matching-based coarsening methods, such as the heavy-edge heuristic, for accelerating spectral clustering [38, 69, 115]. From a theoretical perspective, the approximation quality of matching-based methods was characterized in [91]. Therein, the matching was constructed in the following randomized manner:

Algorithm 5. Randomized edge contraction (one level) [91]
Input. A graph $G = (\mathsf{V}, \mathsf{E})$

1. Associate with each $e_{ij} \in \mathsf{E}$ a probability $p_{ij} > 0$.
2. While $|E| > 0$:
   (a) Draw a sample e_{ij} from E with probability $\propto p_{ij}$.
   (b) Remove from E both e_{ij} as well as all edges sharing a common endpoint with it.
   (c) Construct contraction set (v_i, v_j).

Output: Contraction sets

The following approximation result is known:

Theorem 5.5 (Corollary 5.1 in [91]) *Consider a graph with bounded degrees* $d_i \ll n$ *and* $\lambda_k \leq \min_{e_{ij} \in E} \left\{ \frac{d_i + d_j}{2} \right\}$. *Suppose that the graph is coarsened by Algorithm 5, using a* heavy-edge potential *such that* $p_{ij} \propto w_{ij}$. *For sufficiently large* n*, a single level, and* $\delta > 0$,

$$\left| f(\mathsf{C}^*; \mathsf{X})^{1/2} - f(\tilde{\mathsf{C}}^*; \mathsf{X})^{1/2} \right| = \mathcal{O}\left(\sqrt{\frac{1 - \frac{m}{n}}{\delta} \frac{\sum_{\ell=2}^{k} \lambda_\ell}{\lambda_{k+1} - \lambda_k}} \right)$$

with probability at least $1 - \delta$. *Above,* $\tilde{\mathsf{C}}^*$ *is the optimal k-means solution when using the lifted eigenvectors of* $\mathbf{L}_c$ *as a spectral embedding.*

We deduce that coarsening works better when the spectral clustering problem is easy (as quantified by the weighted gap $\sum_{\ell=2}^{k} \lambda_\ell / (\lambda_{k+1} - \lambda_k)$) and the achieved error is linear on the reduction ratio $1 - m/n$.

There also exist more advanced techniques for selecting contraction sets that come with stronger guarantees w.r.t. the attained reduction and quality of approximation, but feature running time that is not smaller than that of spectral clustering [90]. In particular, these work also with the normalized Laplacian and can be used to achieve multi-level reduction. Roughly, their strategy is to identify and contract sets $\mathsf{S} \subset V$ for which $\mathbf{x}(i) \approx \mathbf{x}(j)$ for all vectors $\mathbf{x} \in \mathbf{U}_k$ and $v_i, v_j \in \mathsf{S}$. This strategy ensures that the best partitionings of G are preserved by coarsening. We will not expand on these methods here as they do not aim to improve the running time of spectral clustering.

5.4.3 *Other Approaches*

In the following, we present two additional approaches for approximately computing spectral embeddings. The former can be interpreted as a sampling-based method (but in a different manner than the techniques discussed so far), whereas the latter is only vaguely linked to sampling. Nevertheless, we find that both techniques are very interesting and merit a brief discussion.

5.4.3.1 Spectral Sparsification

This approach is best suited for cases when the input of spectral clustering is directly a graph.[16] Different from the methods discussed earlier, here the aim is to identify

[16]When one starts from a set of points, it is preferable to sparsify the graph by retaining a constant number of nearest neighbors for each point. The resulting nearest neighbor graph has already $O(n)$ edges, which is the smallest possible.

a Laplacian matrix $\tilde{\mathbf{L}}$ of the same size as $\mathbf{L}$ but with fewer entries. Additionally, it should be ensured that

$$\frac{1}{1+\epsilon}\mathbf{x}^\top \mathbf{L}\mathbf{x} \leq \mathbf{x}^\top \tilde{\mathbf{L}}\mathbf{x} \leq (1+\epsilon)\mathbf{x}^\top \mathbf{L}\mathbf{x} \quad \text{for all} \quad \mathbf{x} \in \mathbb{R}^n \tag{5.19}$$

for some small constant $\epsilon > 0$ [125]. Most fast algorithms for spectral sparsification entail sampling $\mathcal{O}(n \log n)$ edges from the total edges present in the graph. Different sampling schemes are possible [72, 124], but the most popular ones entail sampling edges with replacement based on their effective resistance. It should be noted that though computing all effective resistances exactly can be computationally prohibitive, the effective resistance of edges can be approximated in nearly linear time on the number of edges based on a Johnson–Lindenstrauss argument [124].

There are different ways to use sparsification in order to accelerate spectral clustering. The most direct one is to exploit the fact that the eigenvalues $\tilde{\lambda}_k$ and eigenvectors $\tilde{\mathbf{U}}_k$ of $\tilde{\mathbf{L}}$ approximate, respectively, the eigenvalues and eigenvectors of $\mathbf{L}$ up to multiplicative error. This yields the same flavor of guarantees as in graph coarsening and ensures that the computational complexity of the partial eigen-decomposition will decrease when $e = \omega(n \log n)$. A variation of this idea was considered in [140], though the latter did not provide a complete error and complexity analysis. Alternative approaches are also possible. We refer the interested reader to [136] for a rigorous argument that invokes a Laplacian solver.

Despite these exciting developments, we should mention that the overwhelming majority of graph sparsification algorithms remain in the realm of theory. That is, we are currently not aware of any practical and competitive implementation and thus retain a measure of skepticism with regard to their utility in the setting of spectral clustering.

5.4.3.2 Random Eigenspace Projection

There also exist approaches that do not explicitly rely on sampling. The key starting point here is that, with regard to spectral clustering, one does not need the eigenvectors exactly—any rotation of $\mathbf{U}_k$ suffices (indeed, k-means is an algorithm based on distances and rotations conserve distances). Even more generally, consider $\tilde{\mathbf{U}}_k \in \mathbb{R}^{n \times m}$ with $m \geq k$ and denote:

$$\epsilon = \min_{\mathbf{Q} \in \mathcal{Q}} \|\mathbf{U}_k \mathbf{I}_{k \times m} \mathbf{Q} - \tilde{\mathbf{U}}_k\|_F,$$

where $\mathcal{Q}$ is the space of $m \times m$ unitary matrices and $\mathbf{I}_{k \times m}$ consists of the first k rows of an $m \times m$ identity matrix.

The following lemma (which is a generalization of Lemma 1) shows how ϵ can be used to provide control on the k-means error:

Lemma 2 (Lemma 3.1 in [95])
Let $\tilde{\mathsf{C}}^*$ be the optimal solution of the k-means problem on $\tilde{\mathbf{U}}_k$. It holds that[17]

$$\left| f(\mathsf{C}^*; \mathsf{X})^{1/2} - f(\tilde{\mathsf{C}}^*; \mathsf{X})^{1/2} \right| \leq 2\epsilon. \tag{5.20}$$

There exists (at least) two approaches to efficiently compute $\tilde{\mathbf{U}}_k$ while controlling ϵ [21, 130] (see also related work in [58]). We will consider here a simple variant of the one proposed in [130] and further analyzed in [95]. Let $\mathbf{G} \in \mathbb{R}^{n\times m}$ be a random Gaussian matrix with centered i.i.d. entries, each having variance $\frac{1}{m}$. Furthermore, suppose that we project $\mathbf{G}$ onto span($\mathbf{U}_k$) by multiplying each one of its columns by an ideal projector $\mathbf{P}_k$ defined as

$$\mathbf{P}_k = \mathbf{U} \begin{pmatrix} \mathbf{I}_k & 0 \\ 0 & 0 \end{pmatrix} \mathbf{U}^\top. \tag{5.21}$$

Theorem 5.6 ([95, 130]) *Let $\tilde{\mathsf{C}}^*$ be the optimal solution of the k-means problem on the rows of $\tilde{\mathbf{U}}_k = \mathbf{P}_k\mathbf{G}$. For every $\delta \geq 0$, one has*

$$\left| f(\mathsf{C}^*; \mathsf{X})^{1/2} - f(\tilde{\mathsf{C}}^*; \mathsf{X})^{1/2} \right| \leq 2\sqrt{\frac{k}{m}}\,(\sqrt{k} + \delta), \tag{5.22}$$

with probability at least $1 - \exp(-\delta^2/2)$.

This result means that for an ideal projector $\mathbf{P}_k$, dimension $m = \mathcal{O}(k^2)$ suffices to guarantee good approximation (since the error becomes independent of k and n)! A similar argument also holds when the entries of $\mathbf{G}$, instead of being Gaussian, are selected i.i.d. from $\{-\sqrt{3}, 0, +\sqrt{3}\}$ with probabilities $\{1/6, 2/3, 1/6\}$, respectively [1]. This construction has the benefit of being sparser and, moreover, is reminiscent of sampling. It should be noted that in [130], $m = \mathcal{O}(\log n)$ was deemed enough because one only wanted that the distance between two rows of $\mathbf{U}_k$

---

[17]**A remark on the definition of the k-means cost.** Note that, here, the lines $\tilde{\mathbf{X}}$ of $\tilde{\mathbf{U}}_k$ are points in dimension $m \geq k$, such that the optimal centroid set $\tilde{\mathsf{C}}^*$ minimizing the k-means cost on $\tilde{\mathbf{X}}$ is a set of k points in dimension $m \geq k$. In this context, the notation $f(\tilde{\mathsf{C}}^*; \mathsf{X})$ is ill-defined: it is a sum of distances between points that do not necessarily have the same dimension. We abuse notations and give the following meaning to $f(\tilde{\mathsf{C}}; \mathsf{X})$. First, consider the matrix form of the k-means cost, as used in the proofs of Lemmas 1 and 2: $f(\mathsf{C}; \mathsf{X}) = \|X - \mathbf{C}\mathbf{C}^\top\mathbf{X}\|_F^2$, where $\mathbf{X} = (\mathbf{x}_1|\ldots|\mathbf{x}_n)^\top \in \mathbb{R}^{n\times k}$ and $\mathbf{C} \in \mathbb{R}^{n\times k}$ is the (weighted) cluster indicator matrix associated to the Voronoi tessellation of X given C: $\mathbf{C}_{i\ell} = 1/\sqrt{s_\ell}$ if data point i belongs to cluster ℓ, and 0 otherwise, where s_ℓ is the size of cluster ℓ. Now, let $\tilde{\mathbf{C}} \in \mathbb{R}^{n\times k}$ be the cluster indicator matrix associated with the Voronoi tessellation of $\tilde{\mathsf{X}}$ given $\tilde{\mathsf{C}}$. One writes: $f(\tilde{\mathsf{C}}; \mathsf{X}) = \|\mathbf{X} - \tilde{\mathbf{C}}\tilde{\mathbf{C}}^\top\mathbf{X}\|_F^2$.

was approximated by the distance between the same two rows of $\tilde{\mathbf{U}}_k$. There was in fact no end-to-end control on the k-means error.

The discussion so far assumed that $\mathbf{P}_k$ is an ideal projector onto span($\mathbf{U}_k$). However, in practice one does not have access to this projector as we are in fact in the process of *computing* $\mathbf{U}_k$. One may choose to approximate the action of $\mathbf{P}_k$ by an application of a matrix function h on the representative matrix $\mathbf{R}$ [111, 129]. Assuming a point λ_* in the interval $[\lambda_k, \lambda_{k+1})$ is known, one may select a polynomial [122] or rational function [65, 92] that approximates the ideal low-pass response, i.e., $h(\lambda) = 1$ if $\lambda \leq \lambda_*$ and $h(\lambda) = 0$, otherwise. The approximated projector $\tilde{\mathbf{P}}_k = h(\mathbf{R})$ can be designed to be very close to $\mathbf{P}_k$. For instance, in the case of Chebyshev polynomials of order c using the arguments of [80, Lemma 1] it is easy to prove that w.h.p. using $h(\mathbf{R})$ instead of $\mathbf{P}_k$ does not add more than $\mathcal{O}(c^{-c}\sqrt{n})$ error in (5.20). Furthermore, the operation $\tilde{\mathbf{P}}_k\mathbf{G}$ can conveniently be computed in $\mathcal{O}(mce)$ time via this polynomial approximation.

The last ingredient needed for this approximation is λ_*, i.e., a point in the interval $[\lambda_k, \lambda_{k+1})$. Finding efficiently a valid λ_* is difficult. An option is to rely on eigencount techniques [39, 105, 109] to find one in[18] $\mathcal{O}(ck^2(\log n)(e + n\log(\lambda_n/(\lambda_{k+1} - \lambda_k))))$ time, which features similar complexity as the Lanczos method (see discussion in Sect. 5.2.3). Another option is to content oneself with values of λ_* known only to be close to the interval $[\lambda_k, \lambda_{k+1})$, but thereby loosing the end-to-end guarantee [130].

5.5 Sampling in the Spectral Feature Space

Having computed (or approximated) the spectral embedding $\mathsf{X} = (\mathbf{x}_1, \mathbf{x}_2, \ldots, \mathbf{x}_n)$, what remains is to solve the k-means problem on X, in order to obtain k centroids together with the associated k classes obtained after Voronoi tessellation.

The usual heuristic used to solve the k-means problem, namely the Lloyd-Max algorithm, is already very efficient as it runs in $\mathcal{O}(nk^2t)$ time as seen in Sect. 5.2.3. Nonetheless, this section considers ways to accelerate k-means even further. In the following, we classify the relevant literature in five categories and point towards representative references for each case. In our effort to provide depth (as well as

[18] *Proof sketch*: Given $\lambda \in (0, \lambda_n]$, denote by j the largest integer such that $\lambda_j \leq \lambda$ and by $\mathbf{P}_j$ the orthogonal projector on $\mathbf{U}_j$. Let $\mathbf{G} \in \mathbb{R}^{n\times m'}$ be a random Gaussian matrix with centered i.i.d. entries, each having variance $\frac{1}{m'}$ and denote by $\hat{j} = \|\mathbf{P}_j\mathbf{G}\|_F$. Relying on Theorem 4.1 (and the following discussion in Section 4.2) of [109] with $\mathsf{E}_\lambda = \mathbf{0}$, one has with prob. at least $1 - \epsilon$ that $(1-\delta)j \leq \hat{j} \leq (1+\delta)j$ for all $j = 1, \ldots, n$ provided $m' \geq \frac{1}{\delta^2}\log\frac{n}{\epsilon}$. Setting $\delta = 1/(2k+3)$ gives w.h.p. that $\frac{2k+2}{2k+3}j \leq \hat{j} \leq \frac{2k+4}{2k+3}j$ for all $j = 1, \ldots, n$ provided $m' \geq \mathcal{O}(k^2\log n)$. This implies that w.h.p. for every $j \leq k+1$ it must be that round$(\hat{j}) = j$, whereas when $j > k+1$ we have round$(\hat{j}) > k+1$. Note that round$(\hat{j})$ is the closest integer to $\hat{j}$. By dichotomy on $\lambda \in (0, \lambda_n]$, one thus finds a λ_* in time $\mathcal{O}(ck^2(\log n)(e + n\log(\lambda_n/(\lambda_{k+1} - \lambda_k))))$.

breadth) of presentation, the rest of the section details only methods that belong to the first and last categories.

- **Exact acceleration of Lloyd-Max.** There exists exact accelerated Lloyd-Max algorithms, some of them based on avoiding unnecessary distance calculations using the triangular inequality [59, 101], or on optimized data organization [67], and others concentrating on clever initializations [6, 104]. Unlike the former methods, the latter methods involve sampling and are discussed in Sect. 5.5.1.
- **Approximate acceleration of Lloyd-Max.** Approximately accelerating the Lloyd-Max algorithm has also received attention, for instance, via approximate nearest neighbor methods [108], via cluster closure [142], or via applying Lloyd-Max hierarchically (in the large k context) [103]. An approach involving sampling is introduced in [119]: it is based on mini-batches sampled uniformly at random from $\mathbf{X}$. We will not discuss further this method as it does not come with guarantees on the cost of the obtained solution.
- **Methods involving sampling in the Fourier domain.** There are a few sampling-based heuristics to solve the k-means problem that are different from the Lloyd-Max algorithm. For instance, the work in [70] proposes to sample in the frequency domain to obtain a sketch from which one may recover the centroids with an orthogonal matching pursuit algorithm specifically tailored to this kind of compressive learning task [55]. These methods are reminiscent of the random Fourier features sketching approach introduced in Sect. 5.3.2. We will not discuss them further.
- **Methods involving sampling features.** Similarly to ideas presented in Sect. 5.3.4.2 but here specific to the k-means setting, some works reduce the ambient dimension of the vectors, either by selecting a limited number of features [3, 20] or by embedding all points in a lower dimension using random projections [22, 33, 93]. The tightest results to day are a $(1 + \epsilon)$ multiplicative error on the k-means cost f either by randomly selecting $\mathcal{O}(\epsilon^{-2}k \log k)$ features or by projecting them on a random space of dimension $\mathcal{O}(\epsilon^{-2} \log (k/\epsilon))$ (sublinear in k!). The sampling result is useless in the spectral clustering setting as the ambient dimension of the spectral features is already k. The projection result could in principle be applied in our setting, to reduce the cost of the k-means step to $\mathcal{O}(tnk \log k)$. We will nevertheless not discuss it further in this chapter.
- **Methods involving sampling points.** Finally, the last group of existing methods are the ones that solve k-means on a subset S of X, before lifting back the result on the whole dataset. We classify such methods in two categories. In Sect. 5.5.2, we detail methods that are graph-agnostic, meaning that they apply to *any* k-means problem; and in Sect. 5.5.3 we discuss methods that explicitly rely on the fact that the features $\mathbf{x}$ were in fact obtained from a known graph. We argue that the latter are better suited to the spectral clustering problem.

5.5.1 Clever Initialization of the Lloyd-Max Algorithm

Recall that the k-means objective on X is to find the k centroids $\mathsf{C} = (\mathbf{c}_1, \ldots, \mathbf{c}_k)$ that minimize the following cost function:

$$f(\mathsf{C}; \mathsf{X}) = \sum_{\mathbf{x} \in \mathsf{X}} \min_{\mathbf{c} \in \mathsf{C}} \|\mathbf{x} - \mathbf{c}\|_2^2. \tag{5.23}$$

and that $\mathsf{C}^* = \arg\min_{\mathsf{C}} f(\mathsf{C}; \mathsf{X})$ is the optimal solution attaining cost $f^* = f(\mathsf{C}^*; \mathsf{X})$. Recall also that the Lloyd-Max algorithm (see Algorithm 2) converges to a local minimum of f that we will denote by C_{lm}, for which the cost function equals $f_{\text{lm}} = f(\mathsf{C}_{\text{lm}}; \mathsf{X})$. It is crucial to note that the initialization of centroids C_{ini} in the first step of the Lloyd-Max algorithm, which usually is done by randomly selecting k points in X, is what determines the distance $|f^* - f_{\text{lm}}|$ to the optimal. As such, significant efforts have been devoted to smartly selecting C_{ini} by various sampling schemes.

As usual, we also face here the usual trade-off between sampling *effectively* and *efficiently*. The fastest sampling method is of course uniformly at random, but it does not come with any guarantee on the quality of the local minimum C_{lm} it leads to. An alternative sampling scheme, called k-means++ initialization, is based on the following more general $\mathbf{D}^2$-`sampling` algorithm.

Algorithm 6: $\mathbf{D}^2$-`sampling`
Input. X, m the number of required samples

1. Initialize B with any $\mathbf{x}$ chosen uniformly at random from X.
2. Iterate the following steps until B contains m elements:
   (a) Compute $d_i = \min_{b \in \mathsf{B}} \|\mathbf{x}_i - b\|_2^2$.
   (b) Define the probability of sampling $\mathbf{x}_i$ as $d_i / \sum_i d_i$.
   (c) Sample $\mathbf{x}_{\text{new}}$ from this probability distribution and add it to B.

Output: B a sample set of size m.

k-means++ initialization boils down to running Algorithm 6 with $m = k$ to obtain a set of k initial centroids. Importantly, when the Lloyd-Max heuristic is run with this initialization, the following guarantee holds:

Theorem 5.7 ([6]) *For any set of data points, the cost f_{lm} obtained after Lloyd-Max initialized with k-means++ is controlled in expectation:* $\mathbb{E}(f_{\text{lm}}) \leq 8(\log k + 2) f^*$.

In terms of computation cost, $\mathbf{D}^2$-`sampling` with $m = k$ runs in $\mathcal{O}(nkd)$, that is, $\mathcal{O}(nk^2)$ in our setting of a spectral embedding X in dimension k. This work inspired other initialization techniques that come with similar guarantees and are in some cases faster [10, 12]. The interested reader is referred to the review [27] for further analyses on the initialization of k-means.

5.5.2 *Graph Agnostic Sampling Methods: Coresets*

The rest of Sect. 5.5 considers sampling methods that fall in the following framework: (i) sample a subset S of X, (ii) solve k-means on S, (iii) lift the result back on the whole dataset X. Section 5.5.2 focuses on coresets: general sampling methods designed for any arbitrary k-means problem, whereas in Sect. 5.5.3, we will take into account the specific nature of the spectral features encountered in spectral clustering algorithms.

5.5.2.1 Definition

Let $\mathsf{S} \subset \mathsf{X}$ be a subset of X of size m. To each element $s \in \mathsf{S}$ associate a weight $\omega(s) \in \mathbb{R}^+$. Define the estimated k-means cost associated with the weighted set S as:

$$\tilde{f}(\mathsf{C}; \mathsf{S}) = \sum_{s \in \mathsf{S}} \omega(s) \min_{\mathbf{c} \in \mathsf{C}} \|s - \mathbf{c}\|_2^2. \tag{5.24}$$

Definition 5.2 (Coreset) Let $\epsilon \in (0, \frac{1}{2})$. The weighted subset S is a ϵ-coreset for f on X if, *for every set* C, the estimated cost is equal to the exact cost up to a relative error:

$$\forall \mathsf{C} \qquad \left| \frac{\tilde{f}(\mathsf{C}; \mathsf{S})}{f(\mathsf{C}; \mathsf{X})} - 1 \right| \leq \epsilon. \tag{5.25}$$

This is the so-called strong coreset definition,[19] as the ϵ-approximation is required for all C. The great interest of finding a coreset S comes from the following fact. Writing $\tilde{\mathsf{C}}^*$ the set minimizing $\tilde{f}$, the following inequalities hold:

$$(1-\epsilon) f(\mathsf{C}^*; \mathsf{X}) \leq (1-\epsilon) f(\tilde{\mathsf{C}}^*; \mathsf{X}) \leq \tilde{f}(\tilde{\mathsf{C}}^*; \mathsf{S}) \leq \tilde{f}(\mathsf{C}^*; \mathsf{S}) \leq (1+\epsilon) f(\mathsf{C}^*; \mathsf{X}).$$

[19] A weaker version of this definition exists in the literature where the ϵ-approximation is only required for C^*.

The first inequality comes from the fact that C^* is optimal for f, the second and last inequality are justified by the coreset property of S, and the third inequality comes from the optimality of $\tilde{\mathsf{C}}^*$ for $\tilde{f}$. This has two consequences:

1. First of all, since $\epsilon < \frac{1}{2}$:

$$f(\mathsf{C}^*; \mathsf{X}) \leq f(\tilde{\mathsf{C}}^*; \mathsf{X}) \leq (1 + 4\epsilon) f(\mathsf{C}^*; \mathsf{X}),$$

   meaning that $\tilde{\mathsf{C}}^*$ is a well-controlled approximation of C^* with a multiplicative error on the cost.
2. Estimating $\tilde{\mathsf{C}}^*$ can be done using the Lloyd-Max algorithm on the weighted subset[20] S, thus reducing the computation time from $\mathcal{O}(nk^2)$ to $\mathcal{O}(mk^2)$.

Coreset methods for k-means thus follow the general procedure:

Algorithm 7. Coresets to avoid k-means on X
Input. X, sampling set size m, and number of clusters $k \leq m$.

1. Compute a weighted coreset S of size m using a `coreset-sampling` algorithm.
2. Run the Lloyd-Max algorithm on the weighted set S to obtain the set of k centroids $\tilde{\mathsf{C}}$.
3. "Closest-centroid lifting": classify the whole dataset X based on the Voronoi cells of $\tilde{\mathsf{C}}$.

Output: A set of k centroids $\mathsf{C} = (\mathbf{c}_1, \ldots, \mathbf{c}_k)$.

Coreset methods compete with one another on essentially two levels: the coreset size m should be as small as possible in order to decrease the time of Lloyd-Max on S, and the coreset itself should be sampled efficiently (at least faster than running k-means on the whole dataset!), which turns out in fact to be a strong requirement. The reader interested in an overview of coreset construction techniques is referred to the recent review [99], as well as Chap. 2 of this book.

5.5.2.2 An Instance of Coreset-Sampling Algorithm

We focus on a particular coreset algorithm proposed in [11] that builds upon results developed in [45, 79]: it is not state of the art in terms of coreset size, but has the advantage of being easy to implement and fast enough to compute. It reads:

---

[20]Generalizing Algorithm 2 to a weighted set is straightforward: in step 2b, instead of computing the center of each cluster, compute the weighted barycenter.

Algorithm 8: a coreset sampling algorithm [11]
Input. X, m the number of required samples, t an iteration number

1. Repeat t times: draw a set of size k using $\mathbf{D}^2$-`sampling`. Out of the t sets obtained, keep the set B that minimizes $f(\mathsf{B};\mathsf{X})$.
2. $\alpha \leftarrow 16(\log k + 2)$
3. For each $b_\ell \in \mathsf{B}$, define B_ℓ the set of points in X in the Voronoi cell of b_ℓ
4. Set $\phi = \frac{1}{n} f(\mathsf{B};\mathsf{X})$.
5. For each $b_\ell \in \mathsf{B}$ and each $\mathbf{x} \in \mathsf{B}_\ell$, define

$$s(\mathbf{x}) = \frac{\alpha}{\phi}\|\mathbf{x} - b_\ell\|_2^2 + \frac{2\alpha}{\phi|\mathsf{B}_\ell|}\sum_{\mathbf{x}'\in\mathsf{B}_\ell}\|\mathbf{x}' - b_\ell\|_2^2 + \frac{4n}{|\mathsf{B}_\ell|}$$

6. Define the probability of sampling $\mathbf{x}_i$ as $\mathbf{p}_i = s(\mathbf{x}_i)/\sum_{\mathbf{x}} s(\mathbf{x})$
7. $\mathsf{S} \leftarrow$ sample m nodes i.i.d. with replacement from $\mathbf{p}$ and associate to each sample s the weight $\omega_s = \frac{1}{m p_s}$.

Output: A weighted set S of size m.

Theorem 2.5 of [11] states:

Theorem 5.8 *Let $\epsilon \in (0, 1/4)$ and $\delta \in (0, 1)$. Let S be the output of Algorithm 8 with $t = \mathcal{O}(\log 1/\delta)$. Then, with probability at least $1 - \delta$, S is a ϵ-coreset provided that:*

$$m = \Omega\left(\frac{k^4 \log k + k^2 \log 1/\delta}{\epsilon^2}\right). \tag{5.26}$$

The computation cost of running this coreset-sampling algorithm, running Lloyd-Max on the weighted coreset, and lifting the result back to X is dominated, when[21] $n \gg k$, by step 1 of Algorithm 6 and thus sums up to $\mathcal{O}(nk^2 \log 1/\delta)$.

Remark 1 The coreset-sampling strategy underlying this algorithm relies on the concept of sensitivity [79]. Many other constructions of coresets for k-means are possible [99] with better theoretical bounds then (5.26). Nevertheless, as the coreset line of research has been essentially theoretical, practical implementations of coreset-sampling algorithms are scarce. A notable exception is, for instance, the work in [49] that proposes a scalable hybrid coreset-inspired algorithm for k-means. Other exceptions are the sampling algorithms based on the farthest-first procedure, a variant of $\mathbf{D}^2$-`sampling` that chooses each new sample to be $\arg\max_i d_i$ instead of drawing it according to a probability proportional to d_i. Once S of size m is drawn,

[21] To be precise, the statement holds if $n \geq \mathcal{O}\left(\frac{k^4}{\epsilon^2}\frac{\log k}{\log 1/\delta}\right)$.

then $\forall s \in \mathsf{S}$, each weight ω_s is set to be the cardinal of the Voronoi cell associated with s. Authors in [113] show that such weighted sets computed by different variants of the farthest-first algorithm are ϵ-coresets, but for values of ϵ that can be very large. For a fixed ϵ, the number of samples necessary to have a ϵ-coreset with this type of algorithm is unknown (see also Chap. 3 of this book).

5.5.3 *Graph-Based Sampling Methods*

The methods discussed so far in this section are graph agnostic both for the sampling procedure *and* the lifting: they do not take into account that, in spectral clustering, X are in fact spectral features of a known graph.

A recent line of work [52, 53, 95, 130] based on graph signal processing (GSP) [118, 123] leverages this additional knowledge for accelerating both the sampling and the lifting steps. For the purpose of the following discussion, denote by $\mathbf{z}_\ell \in \mathbb{R}^n$ the ground truth indicator vector of cluster ℓ, i.e., $\mathbf{z}_\ell(i) = 1$ if node v_i is in cluster ℓ, and 0 otherwise. The goal of spectral clustering is, of course, to recover $\{\mathbf{z}_\ell\}_{\ell=1,\dots,k}$.

Broadly, GSP-based methods can be summarized in the following general methodology [130]:

Algorithm 9. Graph-based sampling strategies to avoid k-means on X
Input. X, m the number of required samples, k the number of desired clusters

1. Choose the random sampling strategy. Either:
   (a) **uniform (i.i.d.)** Draw m i.i.d. samples uniformly.
   (b) **leverage score (i.i.d.)** Compute $\forall \mathbf{x}_i, p_i^* = \|\mathbf{U}_k^\top \delta_i\|_2^2/k$. Draw m i.i.d. samples from $\mathbf{p}^*$. (optional:) set the weight of each sample s to $1/p_s^*$.
   (c) **DPP** Sample a few times independently from a DPP with kernel $\mathbf{K}_k = \mathbf{U}_k\mathbf{U}_k^\top$. (optional:) set the weight of each sample s to $1/\pi_s$.
2. Run the Lloyd-Max algorithm on the (possibly weighted) set S to obtain the k reduced cluster indicator vectors $\mathbf{z}_\ell^r \in \mathbb{R}^m$.
3. Lift each reduced indicator vector $\{\mathbf{z}_\ell^r\}_{\ell=1,\dots,k}$ to the full graph either with
   (a) **Least-square** Solve (5.33) with $\mathbf{y} \leftarrow \mathbf{z}_\ell^r$.
   (b) **Tikhonov** Solve (5.34) with $\mathbf{y} \leftarrow \mathbf{z}_\ell^r$.

(continued)

In both cases, $\mathbf{P}_{\mathsf{S}}$ should be set to $\frac{1}{N}\mathbf{I}_m$ if uniform sampling was chosen, to $\mathrm{diag}(p^*_{s_1}, \ldots p^*_{s_m})$ if leverage score sampling was chosen, and to $\mathrm{diag}(\pi_{s_1}, \ldots \pi_{s_m})$ if DPP sampling was chosen.

4. Assign each node j to the cluster ℓ for which $\hat{\mathbf{z}}_\ell(j)/\|\hat{\mathbf{z}}_\ell\|_2$ is maximal.

Output: A partition of X in k clusters

To aid understanding, let us start by a high-level description of Algorithm 9. The indicator vectors $\mathbf{z}_\ell$ are interpreted as graph signals that are (approximately) bandlimited on the similarity graph G (see Sect. 5.5.3.1 for a precise definition). As such, there is no need to measure these indicator vectors everywhere: one can take advantage of generalized Shannon-type sampling theorems to select the set S of m nodes to measure (step 1). Then k-means is performed on S to obtain the indicator vectors $\mathbf{z}^r_\ell \in \mathbb{R}^m$ on the sample set S (step 2). These reduced indicator vectors are interpreted as noisy measurements of the global cluster indicator vectors $\mathbf{z}_\ell$ on S. The solutions $\mathbf{z}^r_\ell$ are lifted back to X as $\hat{\mathbf{z}}_\ell$ via solving an inverse problem taking into account the bandlimitedness assumption or via label-propagation on the graph structure reminiscent of semi-supervised learning techniques (step 3). As the lifted solutions $\hat{\mathbf{z}}_\ell$ do not have a binary structure as true indicator vectors should have, an additional assignment step is necessary: assign each node j to the class ℓ for which $\frac{\hat{\mathbf{z}}_\ell(j)}{\|\hat{\mathbf{z}}_\ell\|_2}$ is maximal (step 4).

The rest of this section is devoted to the discussion of the three sampling schemes as well as the two lifting procedures considered in this framework. To this end, we will first introduce a few graph signal processing (GSP) concepts in Sec. 5.5.3.1 before discussing in Sec. 5.5.3.2 several examples of graph sampling theorems appropriate to the spectral clustering context.

5.5.3.1 A Brief Introduction to Graph Signal Processing (GSP)

Denote by $\mathbf{U} = (\mathbf{u}_1|\ldots|\mathbf{u}_n) \in \mathbb{R}^{n\times n}$ the matrix of orthonormal eigenvectors of the Laplacian matrix $\mathbf{L}$, with the columns ordered according to their associated sorted eigenvalues: $0 = \lambda_1 \leq \lambda_2 \leq \ldots \leq \lambda_n$. In the GSP literature [118, 123], these eigenvectors are interpreted as graph Fourier modes for two main reasons:

- By analogy to the ring graph, whose Laplacian matrix is exactly the (symmetric) double derivative discrete operator, and is thus diagonal in the basis formed by the classical 1D discrete Fourier modes.
- A variational argument stemming from the Dirichlet form can be exploited to express eigenvectors $\mathbf{u}_i$ of $\mathbf{L}$ as the basis of minimal variation $\mathbf{x}^\top\mathbf{L}\mathbf{x} = \frac{1}{2}\sum_{ij}\mathbf{W}_{ij}\,[\mathbf{x}(i) - \mathbf{x}(j)]^2$ on G and eigenvalues λ_i as a sum of local variations of $\mathbf{u}_i$, i.e., a generalized graph frequency.

A *graph signal* $\mathbf{z} \in \mathbb{R}^n$ is a signal that is defined on the nodes of a graph: its i-th element is associated with node v_i. Given the previous discussion, the graph Fourier transform of $\mathbf{z}$, denoted by $\tilde{\mathbf{z}}$, is its projection on the graph Fourier modes: $\tilde{\mathbf{z}} = \mathbf{U}^\top \mathbf{z} \in \mathbb{R}^n$. The notion of graph filtering naturally follows as a multiplication in the Fourier domain. More precisely, define a real-valued filter function $h(\lambda)$ defined on $[0, \lambda_n]$. The signal $\mathbf{x}$ filtered by h reads $\mathbf{U}h(\Lambda)\mathbf{U}^\top\mathbf{x}$, where we use the convention $h(\Lambda) = \text{diag}(h(\lambda_1), h(\lambda_2), \ldots, h(\lambda_n))$. In the following, we will use the following notation for graph filter operators:

$$h(\mathbf{L}) = Uh(\Lambda)\mathbf{U}^\top. \tag{5.27}$$

For more details on the graph Fourier transform and filtering, their various definitions and interpretations, we refer the reader to [133].

Of interest for the discussion in this chapter, one may define bandlimited graph signals as linear combinations of the first few low-frequency Fourier modes. Writing $\mathbf{U}_k = (\mathbf{u}_1|\ldots|\mathbf{u}_k) \in \mathbb{R}^{n\times k}$, we have the formal definition:

Definition 5.3 (*k*-Bandlimited Graph Signal) A graph signal $\mathbf{z} \in \mathbb{R}^n$ is k-bandlimited if $\mathbf{z} \in \text{span}(\mathbf{U}_k)$, i.e., $\exists\, \boldsymbol{\alpha} \in \mathbb{R}^k$ such that $\mathbf{z} = \mathbf{U}_k\boldsymbol{\alpha}$.

To grasp why the notion of k-bandlimitedness lends itself naturally to the approximation of spectral clustering, consider momentarily a graph with k disconnected components and $\mathbf{z}_\ell \in \mathbb{R}^n$ the indicator vector of cluster ℓ. It is a well-known property of the (combinatorial) Laplacian that $\{\mathbf{z}_\ell\}_{\ell=1,\ldots,k}$ form a set of orthogonal eigenvectors of $\mathbf{L}$ associated with eigenvalue 0: that is, the set of indicator vectors $\{\mathbf{z}_\ell\}_{\ell=1,\ldots,k}$ form a basis of $\text{span}(\mathbf{U}_k)$. Understanding arbitrary graphs with block structure as a perturbation of the ideal disconnected component case, the indicator vectors $\{\mathbf{z}_\ell\}_{\ell=1,\ldots,k}$ of the blocks should live close to $\text{span}(\mathbf{U}_k)$ (in the sense that the difference between any $\mathbf{z}_\ell$ and its orthogonal projection onto $\text{span}(\mathbf{U}_k)$ is small). This in turn implies that every $\mathbf{z}_\ell$ should be approximately k-bandlimited.

As we will see next, the bandlimitedness assumption is very useful because it enables us to make use of generalized versions of Nyquist–Shannon sampling theorems, taking into account the graph.

5.5.3.2 Graph Sampling Theorems

The periodic sampling paradigm of the Shannon theorem for classical bandlimited signals does not apply to graphs without specific regular structure. In fact, a number of sampling schemes have been recently developed with the purpose of generalizing sampling theorems to graph signals [30, 109, 117, 134] (see [88] for a review of existing schemes).

Let us introduce some notations. Sampling entails selecting a set $\mathsf{S} = (s_1, \ldots, s_m)$ of m nodes of the graph. To each possible sampling set, we associate a measurement matrix $\mathbf{M} = (\boldsymbol{\delta}_{s_1}|\boldsymbol{\delta}_{s_2}|\ldots|\boldsymbol{\delta}_{s_m})^\top \in \mathbb{R}^{m\times n}$ where $\boldsymbol{\delta}_{s_i}(j) = 1$ if

$j = s_i$, and 0 otherwise. Now, consider a k-bandlimited signal $\mathbf{z} \in \text{span}(\mathbf{U}_k)$. The measurement of $\mathbf{z}$ on S reads:

$$\mathbf{y} = \mathbf{M}\mathbf{z} + \mathbf{n} \in \mathbb{R}^m, \tag{5.28}$$

where $\mathbf{n}$ models measurement noise. The sampling question boils down to: how should we sample S such that one can recover any bandlimited $\mathbf{z}$ given its measurement $\mathbf{y}$? There are three important components to this question: (i) how many samples m do we allow ourselves ($m = k$ being the strict theoretical minimum)? (ii) how much does it cost to sample? (iii) how do we in practice recover $\mathbf{z}$ from $\mathbf{y}$ and how much does that inversion cost?

There are a series of works that propose greedy algorithms to find the "best" set S of minimal size $m = k$ that embed all k-bandlimited signals (see, for instance, [131] and references therein). These algorithms cost $\mathcal{O}(nk^4)$ and are thus not competitive in our setting.[22] Moreover, in our case, we do not really need to be that strict on the number of samples and can allow more than k samples. A better choice is to use random graph sampling techniques. In the following we consider two types of independent sampling (uniform and leverage-score sampling) as well as a more involved method based on determinantal point processes.

Independent Sampling In the i.i.d. setting, one defines a discrete probability distribution $\mathbf{p} \in \mathbb{R}^n$ over the node set V. The sampling set S is then generated by drawing m nodes independently with replacement from $\mathbf{p}$. At each draw, the probability to sample node v_i is denoted by p_i. We have $\sum_i p_i = 1$ and write $\mathbf{P} = \text{diag}(p)$. Under this sampling scheme, the following restricted isometry property holds for the associated measurement matrix $\mathbf{M}$ [109].

Theorem 5.9 *For any* $\delta, \epsilon \in (0, 1)$, *with probability at least* $1 - \delta$*:*

$$(1-\epsilon)\|z\|_2^2 \leq \frac{1}{m}\|\mathbf{M}\mathbf{P}^{-1/2}z\|_2^2 \leq (1+\epsilon)\|z\|_2^2 \tag{5.29}$$

for all $\mathbf{z} \in \text{span}(\mathbf{U}_k)$ *provided that*

$$m \geq \frac{3}{\epsilon^2}(\nu_{\mathbf{p}}^k)^2 \log \frac{2k}{\delta} \tag{5.30}$$

where $\nu_{\mathbf{p}}^k$ *is the so-called graph weighted coherence:*

$$\nu_{\mathbf{p}}^k = \max_i \left\{ p_i^{-1/2} \|\mathbf{U}_k^\top \delta_i\|_2 \right\}. \tag{5.31}$$

This property is important as it says, in a nutshell, that any two different bandlimited signals will be identifiable post-sampling provided the number of samples is large

[22] It takes longer to find a good sample than to run k-means on the whole dataset!

enough. The concept of large enough depends on $(\nu_{\mathbf{p}}^k)^2$: a measure of the interplay between the probability distribution and the norms of the rows of $\mathbf{U}_k$. In the uniform i.i.d. case since $p_i = 1/n$, one has $(\nu_{\mathbf{p}}^k)^2 = n \max_i \|\mathbf{U}_k^\top \delta_i\|_2^2$, which stays under control only for very regular graphs, but can be close to n in irregular graphs such as the star graph. The good news is that there exists an optimal sampling distribution (in the sense that it minimizes the right-hand side of inequality (5.30)) that adapts to the graph at hand:

$$p_i^* = \frac{\|\mathbf{U}_k^\top \delta_i\|_2^2}{k} \tag{5.32}$$

In fact, in this case, $(\nu_{\mathbf{p}^*}^k)^2$ matches its lower bound k and the necessary number of samples m to embed all bandlimited signals drops to $\mathcal{O}(k \log k)$. The distribution $\mathbf{p}^*$ is also referred to by the name "leverage scores" in parts of the literature (see discussion in Sect. 5.3.1) [41]. As such, i.i.d. sampling under $\mathbf{p}^*$ will be referred to as leverage score sampling.

Now, for lifting, there are several options.

- If one uses the unbiased decoder

$$\hat{\mathbf{z}} = \underset{\mathbf{w} \in \text{span}(\mathbf{U}_k)}{\arg\min} \ \|\mathbf{P}_{\mathsf{S}}^{-1/2}(\mathbf{M}\mathbf{w} - \mathbf{y})\|_2^2 \tag{5.33}$$

where $\mathbf{P}_{\mathsf{S}}^{-1/2} = \mathbf{M}\mathbf{P}^{-1/2}\mathbf{M}^\top$, then the following reconstruction result holds [109]:

Theorem 5.10 *Let* S *be the i.i.d. nodes sampled with distribution* $\mathbf{p}$ *and* $\mathbf{M}$ *be the associated sampling matrix. Let* $\epsilon, \delta \in (0, 1)$ *and suppose that* m *satisfies* (5.30). *With probability at least* $1 - \delta$, *for all* $\mathbf{z} \in \text{span}(\mathbf{U}_k)$ *and* $\mathbf{n} \in \mathbb{R}^m$, *the solution* $\hat{\mathbf{z}}$ *of* (5.33) *verifies:*

$$\|\hat{\mathbf{z}} - \mathbf{z}\|_2 \leq \frac{2}{\sqrt{m(1-\epsilon)}} \|\mathbf{P}_{\mathsf{S}}^{-1/2}\mathbf{n}\|_2.$$

This means that a noiseless measurement of a k-bandlimited signal yields a perfect reconstruction. Also, this quantifies how increasing m reduces the error of reconstruction due to a noisy measurement. Note that this error may be large if there is a significant measurement noise on a node that has a low probability of being sampled. However, by definition, this is not likely to happen.

- One can also use a label-propagation decoder reminiscent to semi-supervised learning techniques [15, 28]:

$$\hat{\mathbf{z}} = \underset{\mathbf{w} \in \mathbb{R}^n}{\arg\min} \|\mathbf{P}_{\mathsf{S}}^{-1/2}(\mathbf{M}\mathbf{w} - \mathbf{y})\|_2^2 + \gamma \ \mathbf{w}^\top g(\mathbf{L})\mathbf{w}, \tag{5.34}$$

where γ is a regularization parameter, $g(\mathbf{L})$ a graph filter operator as in (5.27) with $g(\lambda)$ a non-decreasing function. As g is non-decreasing, the regularization term of (5.34) penalizes high frequency solutions, that is, solutions that are not smooth along paths of the graph. Theorems controlling the error of reconstruction are more involved and we refer the reader to Section 3.3 of [109] for details.

- Other decoders [14, 106] are in principle possible, replacing, for instance, the ℓ_2 Laplacian-based regularization $\mathbf{w}^\top g(\mathbf{L})\mathbf{w}$ by ℓ_1-regularizers $\|\nabla \mathbf{w}\|_1$, but they come with an increased computation cost, lesser guarantees, and have not been used for spectral clustering: we will thus not detail them further.

Let us discuss the computation costs of the previous sampling and lifting techniques. In terms of sampling time, uniform sampling is obviously the most efficient and runs in $\mathcal{O}(k)$. Leverage score sampling is dominated by the computation of the optimal sampling distribution $\mathbf{p}^*$ of (5.32), which takes $\mathcal{O}(nk)$ time.[23] In terms of lifting time, solving the decoder of (5.33) costs $\mathcal{O}(nk + mk^2)$. Solving the decoder of (5.34) costs $\mathcal{O}(et)$ via the conjugate gradient method, where t is the iteration number of the gradient solver (usually around 10 or 20 iterations suffice to obtain good accuracy when $g(\mathbf{L}) = \mathbf{L}$).

This discussion calls for a few remarks. First of all, these theorems are valid if we suppose that $\mathbf{z}$ is exactly k-bandlimited, which is in fact only an approximation if we consider $\mathbf{z}$ to be the ground truth indicator vectors of the k clusters to detect in the spectral clustering context. In this case, we can always decompose $\mathbf{z}$ as the sum of its orthogonal projection onto $\text{span}(\mathbf{U}_k)$ and its complement $\boldsymbol{\beta}$: $\mathbf{z} = \mathbf{U}_k\mathbf{U}_k^\top\mathbf{z} + \boldsymbol{\beta}$. (5.28) becomes $\mathbf{y} = \mathbf{M}\mathbf{U}_k\mathbf{U}_k^\top\mathbf{z} + \mathbf{n}$ where $\mathbf{n}$ now represents the sum of a measurement noise and the distance-to-model term $\mathbf{M}\boldsymbol{\beta}$. The aforementioned theorems can then be applied to $\mathbf{U}_k\mathbf{U}_k^\top\mathbf{z}$. Moreover, note that the decoder of (5.34) is not only faster than the other ones in general, it also does not constrain the solution $\hat{\mathbf{z}}$ to be exactly in $\text{span}(\mathbf{U}_k)$, which is in fact desirable in the spectral clustering context: we thus advocate for the decoder of (5.34).

DPP Sampling Determinantal point processes are a class of correlated random sampling strategies that strive to increase "diversity" in the samples, based on a kernel $\mathbf{K}$ expliciting the similarity between variables. DPP sampling has been used successfully in a number of applications in machine learning (see, for instance, [74]).

Denote by $[n]$ the set of all subsets of $\{1, 2, \ldots, n\}$. An element of $[n]$ could be the empty set, all elements of $\{1, 2, \ldots, n\}$ or anything in between. DPPs are defined as follows:

Definition 5.4 (Determinantal Point Process [74]) Consider a point process, i.e., a process that randomly draws an element $\mathsf{S} \in [n]$. It is determinantal if, $\forall\, \mathsf{A} \subseteq \mathsf{S}$,

[23] Note that the complexity is different from the leverage score computation of the Nyström techniques of Sects. 5.3.1 and 5.4.1 because, here, we suppose $\mathbf{U}_k$ known whereas $\mathbf{U}_k$ was not known in the previous sections. With $\mathbf{U}_k$ known, computing the leverage scores only entails computing the normalized energy of each line of $\mathbf{U}_k$.

$$\mathbb{P}(\mathsf{A} \subseteq \mathsf{S}) = \det(\mathbf{K}_{\mathsf{A}}),$$

where $\mathbf{K} \in \mathbb{R}^{n\times n}$, a semi-definite positive matrix $0 \preceq \mathbf{K} \preceq 1$, is called the marginal kernel; and $\mathbf{K}_{\mathsf{A}}$ is the restriction of $\mathbf{K}$ to the rows and columns indexed by the elements of A.

The marginal probability π_i of sampling an element i is thus $\mathbf{K}_{ii}$. Consider the following projective kernel:

$$\mathbf{K}_k = \mathbf{U}_k\mathbf{U}_k^\top. \tag{5.35}$$

One can show that DPP samples from such projective kernels are necessarily of size k. After measuring the k-bandlimited signal $\mathbf{z}$ on a DPP sample S, one has the choice between the same decoders as before (see Eqs. (5.33) and (5.34)). For instance:

Theorem 5.11 *For all* $\mathbf{z} \in \text{span}(\mathbf{U}_k)$*, let* $\mathbf{y} = \mathbf{M}\mathbf{z}+\mathbf{n} \in \mathbb{R}^k$ *be a noisy measurement of* $\mathbf{z}$ *on a DPP sample obtained from kernel* $\mathbf{K}_k$*. The decoder of* (5.33) *with* $\mathbf{P} = \text{diag}(\pi_1, \ldots, \pi_n)$ *necessarily enables perfect reconstruction up to the noise level. Indeed, one obtains:*

$$\|\hat{\mathbf{z}} - \mathbf{z}\|_2 \leq \frac{1}{\sqrt{\lambda_{\min}\left(\mathbf{U}_k^\top\mathbf{M}^\top\mathbf{P}_{\mathsf{S}}^{-1}\mathbf{M}\mathbf{U}_k\right)}}\|\mathbf{P}_{\mathsf{S}}^{-1/2}\mathbf{n}\|_2. \tag{5.36}$$

Proof The proof is only partly in [131] and we complete it here. Let us write $\mathbf{z} = \mathbf{U}_k\alpha$. Solving (5.33) entails computing $\hat{\alpha} \in \mathbb{R}^k$ s.t. $\|\mathbf{P}_{\mathsf{S}}^{-1/2}(\mathbf{M}\mathbf{U}_k\hat{\alpha}-\mathbf{y})\|_2^2$ is minimal. Setting the derivative w.r.t. $\hat{\alpha}$ to 0, and replacing $\mathbf{y}$ by $\mathbf{M}\mathbf{U}_k\alpha + \mathbf{n}$, yields:

$$\mathbf{U}_k^\top\mathbf{M}^\top\mathbf{P}_{\mathsf{S}}^{-1}\mathbf{M}\mathbf{U}_k\hat{\alpha} = \mathbf{U}_k^\top\mathbf{M}^\top\mathbf{P}_{\mathsf{S}}^{-1}\mathbf{M}\mathbf{U}_k\alpha + \mathbf{U}_k^\top\mathbf{M}^\top\mathbf{P}_{\mathsf{S}}^{-1}\mathbf{n}.$$

Recall that S is a sample from a DPP with kernel $\mathbf{K}_k$: $\det(\mathbf{M}\mathbf{U}_k\mathbf{U}_k^\top\mathbf{M}^\top)$ is thus strictly superior to 0, which implies that $\mathbf{M}\mathbf{U}_k$ is invertible, which in turn implies that $\hat{\alpha} = \alpha + (\mathbf{M}\mathbf{U}_k)^{-1}\mathbf{n}$. One thus has $\|\hat{\mathbf{z}} - \mathbf{z}\|_2 = \|\hat{\alpha} - \alpha\|_2 = \left\|(\mathbf{M}\mathbf{U}_k)^{-1}\mathbf{n}\right\|_2 = \left\|(\mathbf{P}_{\mathsf{S}}^{-1/2}\mathbf{M}\mathbf{U}_k)^{-1}\mathbf{P}_{\mathsf{S}}^{-1/2}\mathbf{n}\right\|_2$. Using the matrix 2-norm to bound this error yields

$$\|\hat{\mathbf{z}} - \mathbf{z}\|_2 \leq \sqrt{\lambda_{\max}\left[\left(\mathbf{U}_k^\top\mathbf{M}^\top\mathbf{P}_{\mathsf{S}}^{-1}\mathbf{M}\mathbf{U}_k\right)^{-1}\right]}\|\mathbf{P}_{\mathsf{S}}^{-1/2}\mathbf{n}\|_2,$$

as claimed.

Several comments are in order:

- The particular choice of kernel $\mathbf{K}_k = \mathbf{U}_k\mathbf{U}_k^\top$ implies that the marginal probability of sampling node v_i, $\pi_i = \|\mathbf{U}_k^\top\delta_i\|_2^2$, is proportional to the leverage scores p_i^*. The major difference between the i.i.d. leverage score approach and the DPP approach comes from the negative correlations induced by the DPP. In fact, the probability of jointly sampling nodes v_i and v_j in the DPP case is $\pi_i\pi_j - \mathbf{K}_{ij}^2 = \pi_i\pi_j - (\delta_i^\top \mathbf{U}_k\mathbf{U}_k^\top \delta_j)^2$. The interaction term $(\delta_i^\top \mathbf{U}_k\mathbf{U}_k^\top \delta_j)^2$ will be typically large if v_i and v_j are in the same cluster, and small if not. In other words, different from the i.i.d. leverage score case where each new sample is drawn regardless of the past, the DPP procedure avoids to sample nodes containing redundant information.

- Whereas the leverage score approach only guarantees a RIP with high probability after $\mathcal{O}(k \log k)$ samples, the DPP approach has a stronger deterministic guarantee: it enables perfect invertibility (up to the noise level) after precisely $m = k$ samples. The reconstruction guarantee of (5.36) is nevertheless not satisfactory: even corrected by the marginal probabilities $\mathbf{P}_\mathsf{S}$, the matrix $\mathbf{U}_k^\top \mathbf{M}^\top \mathbf{P}_\mathsf{S}^{-1}\mathbf{M}\mathbf{U}_k$ can still have a very small $\lambda_{\min}$, such that reconstruction may be quite sensitive to noise. Improving this control is still an open problem. In practice, sampling independently 2 or 3 times from a DPP with kernel $\mathbf{K}_k$ creates a set S of size $2k$ or $3k$ that is naturally more robust to noise.
- Whereas independent sampling is straightforward, sampling from a DPP with arbitrary kernel costs in general $\mathcal{O}(n^3)$ (see Algorithm 1 of [74] due to [62]). Thankfully, in the case of a projective kernel such as $\mathbf{K}_k$, one can sample a set in $\mathcal{O}(nk^2)$ based on Algorithm 3 of [132].

5.6 Perspectives

Almost two decades have passed since spectral clustering was first introduced. Since then, a large body of work has attempted to accelerate its computation. *So, has the problem been satisfactorily addressed?*—or, despite all these works, is there still room for improvement and further research?

To answer, we must first define what "satisfactorily addressed" would entail. As we have seen, the prototypical spectral clustering algorithm can be divided into three sub-problems: the similarity graph computation runs in $\mathcal{O}(dn^2)$; the spectral embedding computation runs in $\mathcal{O}(t(ek + nk^2))$ using an Arnoldi algorithm with t implicit restarts and assuming that e is the number of edges; and the k-means step runs in $\mathcal{O}(tnk^2)$, with t now being a bound on the number of iterations of the Lloyd-Max algorithm. Our criteria for evaluating an approximation algorithm aiming to accelerate one (or more) of these sub-problems are two-fold:

- We ask that the approximation algorithm's computation cost is effectively lighter than the cost of the sub-problem(s) it is supposed to accelerate! The ultimate achievement is an order-of-magnitude improvement w.r.t. n (or e), d and/or k, especially when the complexity has no hidden constants (i.e., the algorithm is practically implementable). When such a gain is not possible, a gain on the constants of the theoretical cost is also considered worthwhile.
- The algorithm should come with convincing guarantees in terms of the quality of the found solution. Heuristics or partially motivated methods do not cut it. We

require that, under *mild assumptions*, the proposed solution is *provably close to* the exact solution. Let us clarify two aspects of this statement further:

- It is difficult to concretely classify assumptions as mild, but a useful rule of thumb is checking whether the theoretical results are meaningful for the significant majority of cases where spectral clustering would be used.
- The control of the approximation error comes in different flavors, that we detail here from the tightest to the loosest. The best possible error control in our context is a control over the clustering solution itself, via error measures such as the misclustering rate. This is unfortunately unrealistic in many cases. An excellent alternative is the multiplicative error—considered as the gold standard in approximation theory—over the k-means cost,[24] ensuring that the cost of the approximation is not larger than $1+\epsilon$ times the cost of the exact solution. Next comes the additive error over the cost: ensuring that the cost difference between approximated and exact solutions is not larger than ϵ. All these error controls are referred to as end-to-end controls, and represent the limit of what we will consider a *satisfactory* error control.

Reviewing the literature, we were surprised to discover that there are rarely any algorithms meeting fully the proposed criteria: a faster algorithm with end-to-end control over the approximation error under mild assumptions. Let us revisit one by one the different approaches presented in Sects. 5.3, 5.4, and 5.5 examining them in light of our criteria for success. In each category of approximation algorithms, we order the methods according to the power of their error control.

Sampling Methods in the Original Feature Space [Sect. 5.3]

- *Representative points methods* as described in [63, 147] allow for an end-to-end control on the miss-clustering rate ρ, which is unfortunately quite loose. The constants involved in Theorem 5.3 are in fact undefined—thus potentially large—which is problematic knowing that ρ is by definition between 0 and 1. Also, the theorem's assumptions include independence of the ϵ_i, which is hard to justify in practice. On the other hand, the computation gain of such methods is very appealing.
- *Feature projection methods*, where the dimension d of the original feature space is reduced to a dimension $d' \leq d$ based on Johnson–Lindenstrauss arguments, come with a multiplicative error control on the pairwise distances in the original feature space, thus providing a control on the obtained kernel matrix. The impact of this initial approximation on the final clustering result has not been studied.
- *Nyström-inspired methods* [19, 48, 85, 97] can be very efficient in practice especially because they do not need to build the graph. However, precisely because they do not build the graph, these methods cannot exactly perform two key

[24]A control in terms of the k-means cost is usually considered as k-means is the last step of spectral clustering. Nevertheless, recalling the minimum cut perspective of Sect. 5.2.2, the control should arguably be in terms of `rcut` or `ncut` costs.

parts of the prototypical spectral clustering algorithm: the k-NN sparsification and the exact degree computation. The partial knowledge and sparsity of the kernel matrix also makes sampling difficult, as using leverage scores sampling is not possible anymore, whereas most other sampling schemes do not work very well with sparse matrices and come with weak guarantees. To the extent of our knowledge, there is also no convincing mathematical argument proving that using these methods will yield a clustering that is of similar quality to that produced by the exact spectral clustering algorithm.

- *Sketching methods* such as the random Fourier features [110] is yet another way of obtaining a pointwise multiplicative $(1 + \epsilon)$ error on the Gaussian kernel computation. RFF enable to compute a provably good low-rank approximation of the kernel. They nevertheless suffer from the same problems as Nyström-based techniques: without building the graph, sparsification and degree-normalization are uncontrolled. In addition, the guarantees on the low-rank approximation of the kernel do not transfer easily to guarantees of approximation of the spectral embedding $\mathbf{U}_k$.
- *Approximate nearest neighbors methods* are numerous and varied, and come with different levels of guarantees. Practical implementations of algorithms, however, often set aside theoretical guarantees to gain on efficiency and performances; and comparisons are usually done on benchmarks rather than on theoretical performances. In the best of cases, there is a control on how close the obtained nearest neighbor similarity graph is to the exact one, but with no end-to-end control.

Spectral Embedding Approximation Methods [Sect. 5.4]

- *Random eigenspace projection* is a very fast method and has been rigorously analyzed [95, 105, 111, 130]. It is true that a successful application depends on obtaining a good estimate of the k-th eigenvalue, which is very hard when the k-th eigenvalue gap is relatively small. Nevertheless, our current understanding of spectral clustering suggests that it only works well when the gap is (at least) moderately large. As such, though there are definitely situations in which random eigenspace projection will fail to provide an acceleration, these correspond to cases where one should not be using spectral clustering in the first place. The same argumentation can also be used in defense of all methods that come with mild gap assumptions (see coarsening and spectral sparsification).
- *Simple coarsening methods*, such as the heavy-edge matching heuristic [69], have nearly linear complexity, seem to work well in practice, and are accompanied by end-to-end additive error control [91]. Nevertheless, the current analysis of these heuristics only accounts for very moderate reductions ($m \geq n/2$) and thus does not fully prove their success: in real implementations coarsening is used in a multi-level fashion resulting to a drastic decrease in the graph size ($m = \mathcal{O}(n/2^c)$ for c levels), whereas the end-to-end control only works for a single level.
- *Advanced coarsening methods*, such as local variation methods [90], come with much stronger guarantees that allow for drastic size reduction and acceleration. Yet, thus far, all evidence suggests that finding a good enough coarsening is

computationally as hard as solving the spectral clustering problem itself. As a consequence, it is at this point unclear whether these methods can be used to accelerate spectral clustering.

- *Spectral sparsification techniques* come with excellent guarantees in theory: one may prove that a spectral sparsifier can be computed in nearly linear time and, moreover, the latter's spectrum will be provably close to the original one. Yet, we have reasons to doubt their practicality. Indeed, current algorithms are very complex, feature impractically large constants, and are only relevant for dense graphs. In addition, spectral sparsifiers, by definition, approximate the entire spectrum of a graph Laplacian matrix. However, spectral clustering only needs an approximation of a tiny fraction of the spectrum. From that perspective, it is reasonable to conclude that without modification current approaches will not yield the best possible approximation.
- *Nyström-approximation* applied directly to the Laplacian matrix is a good option, especially when combined with leverage score sampling. Nevertheless, an end-to-end error control has only been partially derived and is not yet satisfactory.

Sampling to Accelerate the k-Means Step [Sect. 5.5]

- *Exact methods to accelerate the Lloyd-Max algorithm*, may they be via avoiding unnecessary distance calculations or via a careful initialization are always useful and should be taken into account.
- *Coresets* come with the strongest guarantees: the minimum number of samples to guarantee a $(1+\epsilon)$ multiplicative error on the cost function has been well studied. Nevertheless, practical coreset-sampling methods are scarce; and in the best cases, the sampling cost is of the same order of the Lloyd-Max running cost itself.
- *Graph-based sampling* comes with strong guarantees, but not over the k-means cost: on the reconstruction error based on a k-bandlimited model that is only an approximation in practice. Moreover, we interpret the reduced indicator vectors $\mathbf{z}_\ell^r$ obtained by running Lloyd-Max on the sampled set S as (possibly noisy) measurements of $\mathbf{z}_\ell$ on S. This interpretation currently lacks solid theoretical ground and impedes an end-to-end control of this approximation method. Nevertheless, the leverage-score-based sampling allows for a reduction in order of magnitude of the Lloyd-Max running cost.
- *Other methods* to accelerate k-means are not always appropriate to the spectral clustering context. Spectral feature dimension reduction is unnecessary in our context where $d = k$, sketching methods appropriate to distributed cases where n is very large are not appropriate neither as the spectral features need centralized data to be computed in any case.

In Practice The attentive reader will have remarked that, unsurprisingly, the tighter the error control, the more expensive the computation, and vice versa. Also, although we have put here an emphasis on the approximation error controls, it should not undermine the fact that methods from the whole spectrum are in practice useful, depending on the situation at hand, and specifically on the range of values of

n, d, and k. In very large d situations, a first step of random projection (or feature selection if some features are suspected to be too noisy) should be considered. Then, in situations where the exact computation of the proximity graph is too expensive, one may resort either to sketching methods or to Nyström-type methods to decrease the cost from quadratic to linear in n, and directly obtain an approximation of the spectral embedding without any explicit graph construction. These methods, however, do not take into account a sparsity constraint on the proximity graph and are usually rough on the degree correction they make.

The role of the sparsity constraint is not well understood theoretically, but seems to be important in some practical cases [138]. In such instances, a better option is to use approximate nearest neighbors methods to create a sparse similarity graph, and work from there. In extremely large data, say $n \geq 10^8$, the only workable methods are the representative-based, with, if possible, a first k-means (or compressive k-means [70]) to reduce n to m, or, in last resort, a uniform random sampling strategy.

In situations where one has to deal with such a large similarity graph that Arnoldi iterations are too expensive to compute the spectral embedding (either a graph created via approximate nearest neighbors or if the original data *is* a graph), projection methods such as in [21, 130], coarsening methods such as in [90], or Nyström-based methods are different possibilities.

Sampling methods to accelerate the last k-means step may seem to be a theoretical endeavor given that the Lloyd-Max algorithm is already very efficient. Due to the quadratic term in k, it is nevertheless in practice useful when k grows large. In this situation, hierarchical k-means [103] is a nice option. Coresets, because they are so stringent on the error control, have a hard time actually accelerating k-means, unless hybrid coreset-inspired methods are envisioned [49]. Finally, graph-based methods, because they take into account that spectral features are in fact derived from the graph itself, enable significant acceleration and are well-suited to the spectral clustering context.

Future Research Different directions of research could be envisioned to improve the state of the art:

- For Nyström-inspired methods in the context of Sect. 5.3 (directly applied on the original data) as well as the other methods based on computing a low-rank approximation of the kernel matrix $\mathbf{K}$, further work is needed to control both the sparsification and the degree correction, in order to bridge the gap between a provably good low-rank approximation of $\mathbf{K}$ to a provably good low-rank approximation of $\mathbf{R}$.
- For Nyström methods in the context of Sect. 5.4 (applied on a known or well-approximated similarity graph), it would be interesting to extend Theorem 5.2 (for instance) to a control over $\|\mathbf{A}_k - \tilde{\mathbf{A}}_k\|$ instead of $\|\mathbf{A} - \tilde{\mathbf{A}}\|$. This would enable a tighter use of Davis–Kahan's perturbation theorem in the discussion of Sect. 5.4.1 and, *in fine*, a better end-to-end guarantee.
- Projection-based methods of Sect. 5.4.3.2 currently necessitate to compute a value λ_* known to be in the interval $[\lambda_k, \lambda_{k+1})$. The algorithm used to do so is based on eigencount techniques that turn out to require as much computation time

as the Lanczos iterations needed to compute $\mathbf{U}_k$ exactly. One should relax this constraint to obtain end-to-end guarantees as a function of the distance between a coarsely estimated λ_* and the target interval.
- The derivation and analysis of randomized multi-level coarsening schemes with end-to-end guarantees is very much an open problem. We suspect that by utilizing spectrum-dependent sampling-schemes akin to leverage-scores one should be able to achieve results superior to heavy-edge matching in nearly linear time.
- There is an interesting similarity between coreset techniques and the graph-based sampling strategies discussed in Sect. 5.5.3 and it would be interesting to investigate this link theoretically, maybe paving the way to coresets for spectral clustering?

Finally, accelerating the prototypical spectral algorithm depicted in Algorithm 1 should not be the sole objective of researchers in this field. Indeed, taking the graph cut point-of-view of Sect. 5.2.2, Algorithm 1 makes three insufficiently motivated choices: (i) To begin with, the sparsification step in Algorithm 1 is not well understood. Apart from the fact that it is always computationally more convenient to work with a sparse similarity graph then a dense one, the precise effect of sparsification on the clustering performance has not been analyzed. (ii) As mentioned in Sect. 5.2.2.3, the relaxation employed by spectral relaxation is not unique. Why should we focus our attention on this one versus another? See, for instance, [24, 112] for recent alternative options. (iii) Finally, the use of k-means on the spectral features is not yet fully justified. Most of the end-to-end guarantees presented here compare the k-means cost of the exact solution to the k-means cost of the approximate solution. Given that the very use of k-means is not theoretically grounded, this choice of guarantee is debatable. Other options such as a control over the `rcut` or `ncut` objectives are possible (as in [96]) and should be further investigated.

Acknowledgements This work was partially supported by the CNRS PEPS I3A (Project RW4SPEC) and the Swiss National Science Foundation (Project DLEG, grant PZ00P2 179981).

References

1. Achlioptas, D.: Database-friendly random projections: Johnson-Lindenstrauss with binary coins. J. Comput. Syst. Sci. **66**(4), 671–687 (2003)
2. Ali, H.T., Couillet, R.: Improved spectral community detection in large heterogeneous networks. J. Mach. Learn. Res. 18(225), 1–49 (2018)
3. Altschuler, J., Bhaskara, A., Fu, G., Mirrokni, V., Rostamizadeh, A., Zadimoghaddam, M.: Greedy column subset selection: new bounds and distributed algorithms. In: Balcan, M.F., Weinberger, K.Q. (eds.) Proceedings of the 33rd International Conference on Machine Learning. Proceedings of Machine Learning Research, vol. 48 , pp. 2539–2548, New York (2016)

4. Anagnostopoulos, E., Emiris, I.Z., Psarros, I.: Low-quality dimension reduction and high-dimensional approximate nearest neighbor. In: Arge, L., Pach, J. (eds.) 31st International Symposium on Computational Geometry (SoCG 2015). Leibniz International Proceedings in Informatics (LIPIcs), Dagstuhl, vol. 34, pp. 436–450 (2015). Schloss Dagstuhl–Leibniz-Zentrum fuer Informatik
5. Andoni, A., Indyk, P.: Near-optimal hashing algorithms for approximate nearest neighbor in high dimensions. In: 2006 47th Annual IEEE Symposium on Foundations of Computer Science (FOCS'06) (2006)
6. Arthur, D., Vassilvitskii, S.: k-means++: the advantages of careful seeding. In: Proceedings of the Eighteenth Annual ACM-SIAM Symposium on Discrete Algorithms, pp. 1027–1035. Society for Industrial and Applied Mathematics, Philadelphia (2007)
7. Arya, S., Mount, D.M., Netanyahu, D.M., Silverman, R., Wu, A.Y.: An optimal algorithm for approximate nearest neighbor searching fixed dimensions. J. ACM **45**(6), 891–923 (1998)
8. Aumüller, M., Bernhardsson, E., Faithfull, A.: ANN-benchmarks: a benchmarking tool for approximate nearest neighbor algorithms. In: Beecks, C., Borutta, F., Kröger, P., Seidl, T. (eds.) Similarity Search and Applications, Cham, pp. 34–49. Springer, Berlin (2017)
9. Avrithis, Y., Emiris, I.Z., Samaras, I.Z.: High-dimensional approximate nearest neighbor: k-d generalized randomized forests. arXiv:1603.09596 [cs] (2016)
10. Bachem, O., Lucic, M., Hassani, H., Krause, A.: Fast and provably good seedings for k-means. In: Lee, D.D., Sugiyama, M., Luxburg, U.V., Guyon, I., Garnett, R. (eds.) Advances in Neural Information Processing Systems, vol. 29, pp. 55–63. Curran Associates, New York (2016)
11. Bachem, O., Lucic, M., Krause, A.: Practical coreset constructions for machine learning. arXiv:1703.06476 [stat] (2017)
12. Bahmani, B., Moseley, B., Vattani, A., Kumar, R., Vassilvitskii, S.: Scalable K-means++. Proc. VLDB Endow. **5**(7), 622–633 (2012)
13. Bai, Z., Demmel, J., Dongarra, J., Ruhe, A., van der Vorst, H.: Templates for the Solution of Algebraic Eigenvalue Problems: A Practical Guide. SIAM, Philadelphia (2000)
14. Bellec, P.C., Salmon, J., Vaiter, S.: A sharp oracle inequality for graph-slope. Electron. J. Stat. **11**(2), 4851–4870 (2017)
15. Bengio, Y., Delalleau, O., Roux, N.L.: Label propagation and quadratic criterion. In: Semi-Supervised Learning, pp. 193–216. MIT Press, Cambridge (2006)
16. Bhatia, R.: Matrix Analysis, vol. 169. Springer Science & Business Media, New York (1997)
17. Bie, T.D., Cristianini, N.: Fast SDP relaxations of graph cut clustering, transduction, and other combinatorial problems. J. Mach. Learn. Res. **7**, 1409–1436 (2006)
18. Bordenave, C., Lelarge, M., Massoulié, L.: Non-backtracking spectrum of random graphs: community detection and non-regular ramanujan graphs. In: 2015 IEEE 56th Annual Symposium on Foundations of Computer Science, pp. 1347–1357 (2015)
19. Bouneffouf, D., Birol, I.: Sampling with minimum sum of squared similarities for Nystrom-based large scale spectral clustering. In: The International Joint Conference on Artificial Intelligence (IJCAI 2015), pp. 2313–2319 (2015)
20. Boutsidis, C., Drineas, P., Mahoney, M.W.: Unsupervised feature selection for the k-means clustering problem. In: Bengio, Y., Schuurmans, D., Lafferty, J.D., Williams, C.K.I., Culotta, A. (eds.) Advances in Neural Information Processing Systems, vol. 22, pp. 153–161. Curran Associates, New York (2009)
21. Boutsidis, C., Gittens, A., Kambadur, P.: Spectral clustering via the power method-provably. In: International Conference on Machine Learning (ICML) (2015)
22. Boutsidis, C., Zouzias, A., Mahoney, M.W., Drineas, P.: Randomized dimensionality reduction for k -means clustering. IEEE Trans. Inf. Theory **61**(2), 1045–1062 (2015)
23. Brand, M., Huang, K.: A unifying theorem for spectral embedding and clustering. In: 9th International Conference on Artificial Intelligence and Statistics, AISTATS (2003)

24. Bresson, X., Laurent, T., Uminsky, D., von Brecht, J.: Multiclass total variation clustering. In: Burges, C.J.C., Bottou, L., Welling, M., Ghahramani, Z., Weinberger, K.Q. (eds.) Advances in Neural Information Processing Systems, vol. 26, pp. 1421–1429. Curran Associates, New York (2013)
25. Cai, D., Zhang, C., He, X.: Unsupervised feature selection for multi-cluster data. In: Proceedings of the 16th ACM SIGKDD International Conference on Knowledge Discovery and Data Mining - KDD'10, Washington, p. 333. ACM Press, New York (2010)
26. Calvetti, D., Reichel, L., Sorensen, D.C.: An implicitly restarted Lanczos method for large symmetric eigenvalue problems. Electron. Trans. Numer. Anal. **2**(1), 21 (1994)
27. Celebi, M.E., Kingravi, H.A., Vela, P.A.: A comparative study of efficient initialization methods for the k-means clustering algorithm. Expert Syst. Appl. **40**(1), 200–210 (2013)
28. Chapelle, O., Schlkopf, B., Zien, A.: Semi-Supervised Learning, 1st edn. MIT Press, Cambridge (2010)
29. Chen, X., Cai, D.: Large scale spectral clustering with landmark-based representation. In: AAAI Conference on Artificial Intelligence (2011)
30. Chen, S., Varma, R., Sandryhaila, A., Kovacevic, J.: Discrete signal processing on graphs: sampling theory. CoRR, abs/1503.05432 (2015)
31. Chitta, R., Jin, R., Jain, A.K.: Efficient kernel clustering using random Fourier features. In: 2012 IEEE 12th International Conference on Data Mining, pp. 161–170 (2012)
32. Chung, F.R., Graham, F.C.: Spectral Graph Theory, vol. 92. American Mathematical Society, Providence (1997)
33. Cohen, M.B., Elder, S., Musco, C., Musco, C., Persu, M.: Dimensionality reduction for k-means clustering and low rank approximation. In: Proceedings of the Forty-Seventh Annual ACM Symposium on Theory of Computing, STOC'15, New York, pp. 163–172. ACM. New York (2015)
34. Dall'Amico, L., Couillet, R., Tremblay, N.: Revisiting the Bethe-Hessian: improved community detection in sparse heterogeneous graphs. In: NeurIPS (2019)
35. Dash, M., Koot, P.W.: Feature selection for clustering. In: Liu, L., özsu, M.T. (eds.) Encyclopedia of Database Systems, pp. 1119–1125. Springer, Boston (2009)
36. Davis, C.: The rotation of eigenvectors by a perturbation. J. Math. Anal. Appl. **6**(2), 159–173 (1963)
37. Deshpande, A., Rademacher, L., Vempala, S., Wang, G.: Matrix approximation and projective clustering via volume sampling. In: Proceedings of the Seventeenth Annual ACM-SIAM Symposium on Discrete Algorithm, pp. 1117–1126. Society for Industrial and Applied Mathematics, Philadelphia (2006)
38. Dhillon, I.S., Guan, Y., Kulis, B.: Weighted graph cuts without eigenvectors a multilevel approach. IEEE Trans. Pattern Anal. Mach. Intell. **29**(11), 1944–1957 (2007)
39. Di Napoli, E., Polizzi, E., Saad, Y.: Efficient estimation of eigenvalue counts in an interval. Numer. Linear Algebra Appl. **23**(4), 674–692 (2016)
40. Dong, W., Moses, C., Li, K.: Efficient k-nearest neighbor graph construction for generic similarity measures. In: Proceedings of the 20th International Conference on World Wide Web - WWW'11, Hyderabad, p. 577. ACM Press, New York (2011)
41. Drineas, P., Mahoney, M.W.: Lectures on randomized numerical linear algebra. Math. Data **25**, 1 (2018)
42. Drineas, P., Frieze, A.M., Kannan, R., Vempala, S., Vinay, V.: Clustering in large graphs and matrices. In: Proceedings of the Tenth Annual ACM-SIAM Symposium on Discrete Algorithms (SODA), vol. 99, pp. 291–299. Citeseer (1999)
43. Drineas, P., Kannan, R., Mahoney, M.W.: Fast Monte Carlo algorithms for matrices ii: computing a low-rank approximation to a matrix. SIAM J. Comput. **36**(1), 158–183 (2006)
44. Duan, R., Pettie, S.: Linear-time approximation for maximum weight matching. J. ACM **61**(1), 1 (2014)
45. Feldman, D., Langberg, M.: A unified framework for approximating and clustering data. In: Proceedings of the Forty-Third Annual ACM Symposium on Theory of Computing, pp. 569–578. ACM, New York (2011)

46. Fiedler, M.: Algebraic connectivity of graphs. Czechoslov. Math. J. **23**(2), 298–305 (1973)
47. Fortunato, S.: Community detection in graphs. Phys. Rep. **486**(3–5), 75–174 (2010)
48. Fowlkes, C., Belongie, S., Chung, F., Malik, J.: Spectral grouping using the Nystrom method. IEEE Trans. Pattern Anal. Mach. Intell. **26**(2), 214–225 (2004)
49. Frahling, G., Sohler, C.: A fast k-means implementations using coresets. Int. J. Comput. Geom. Appl. **18**(6), 605–625 (2008)
50. Frieze, A., Kannan, R., Vempala, S.: Fast Monte-Carlo algorithms for finding low-rank approximations. J. ACM **51**(6), 1025–1041 (2004)
51. Fu, C., Xiang, C., Wang, C., Cai, D.: Fast approximate nearest neighbor search with the navigating spreading-out graph. arXiv:1707.00143 [cs] (2017)
52. Gadde, A., Anis, A., Ortega, A.: Active semi-supervised learning using sampling theory for graph signals. In: Proceedings of the 20th ACM SIGKDD International Conference on Knowledge Discovery and Data Mining, KDD'14, New York, pp. 492–501. ACM, New York (2014)
53. Gadde, A., Gad, E.E., Avestimehr, S., Ortega, A.: Active learning for community detection in stochastic block models. In: 2016 IEEE International Symposium on Information Theory (ISIT), pp. 1889–1893 (2016)
54. Gittens, A., Mahoney, M.W.: Revisiting the Nyström method for improved large-scale machine learning. J. Mach. Learn. Res. **17**(1), 3977–4041 (2016)
55. Gribonval, R., Blanchard, G., Keriven, N., Traonmilin, Y.: Compressive statistical learning with random feature moments. arXiv:1706.07180 [cs, math, stat] (2017)
56. Guattery, S., Miller, G.L.: On the performance of spectral graph partitioning methods. In: Proceedings of the Sixth Annual ACM-SIAM Symposium on Discrete Algorithms SODA, vol. 95, pp. 233–242 (1995)
57. Guattery, S., Miller, G.L.: On the quality of spectral separators. SIAM J. Matrix Anal. Appl. **19**(3), 701–719 (1998)
58. Halko, N., Martinsson, P., Tropp, J.: Finding structure with randomness: probabilistic algorithms for constructing approximate matrix decompositions. SIAM Rev. **53**(2), 217–288 (2011)
59. Hamerly, G., Drake, J.: Accelerating Lloyd's Algorithm for k-Means Clustering, pp. 41–78. Springer, Cham (2015)
60. He, X., Cai, D., Niyogi, P.: Laplacian score for feature selection. In: Advances in Neural Information Processing Systems, pp. 507–514 (2006)
61. Hendrickson, B., Leland, R.W.: A multi-level algorithm for partitioning graphs. Supercomputing **95**(28), 1–14 (1995)
62. Hough, J.B., Krishnapur, M., Peres, Y., Virág, B.: Determinantal processes and independence. Probab. Surv. **3**, 206–229 (2006)
63. Huang, L., Yan, D., Taft, N., Jordan, M.I.: Spectral clustering with perturbed data. In: Koller, D., Schuurmans, D., Bengio, Y., Bottou, L. (eds.) Advances in Neural Information Processing Systems, vol. 21, pp. 705–712. Curran Associates, New York (2009)
64. Hunter, B., Strohmer, T., Simos, T.E., Psihoyios, G., Tsitouras, C.: Compressive spectral clustering, pp. 1720–1722, Rhodes (2010)
65. Isufi, E., Loukas, A., Simonetto, A., Leus, G.: Autoregressive moving average graph filtering. IEEE Trans. Signal Process. **65**(2), 274–288 (2017)
66. Kalantidis, Y., Avrithis, Y.: Locally optimized product quantization for approximate nearest neighbor search. In: The IEEE Conference on Computer Vision and Pattern Recognition (CVPR) (2014)
67. Kanungo, T., Mount, D.M., Netanyahu, N.S., Piatko, C.D., Silverman, R., Wu, A.Y.: An efficient k-means clustering algorithm: analysis and implementation. IEEE Trans. Pattern Anal. Mach. Intell. **24**, 881–892 (2002)
68. Karger, D.R.: Minimum cuts in near-linear time. J. ACM **47**(1), 46–76 (2000)
69. Karypis, G., Kumar, V.: A fast and high quality multilevel scheme for partitioning irregular graphs. SIAM J. Sci. Comput. **20**(1), 359–392 (1998)

70. Keriven, N., Tremblay, N., Traonmilin, Y., Gribonval, R.: Compressive K-means. In: 2017 IEEE International Conference on Acoustics, Speech and Signal Processing (ICASSP), pp. 6369–6373 (2017)
71. Kolev, P., Mehlhorn, K.: A note on spectral clustering. arXiv preprint arXiv:1509.09188 (2015)
72. Koutis, I., Miller, G.L., Peng, R.: Approaching optimality for solving SDD linear systems. In: 2010 IEEE 51st Annual Symposium on Foundations of Computer Science, pp. 235–244 (2010)
73. Krzakala, F., Moore, C., Mossel, E., Neeman, J., Sly, A., Zdeborová, L., Zhang, P.: Spectral redemption in clustering sparse networks. Proc. Natl. Acad. Sci. **110**(52), 20935–20940 (2013)
74. Kulesza, A., Taskar, B.: Determinantal point processes for machine learning. Found. Trends Mach. Learn. **5**(2–3), 123–286 (2012)
75. Kumar, A., Kannan, R.: Clustering with spectral norm and the k-means algorithm. In: 2010 51st Annual IEEE Symposium on Foundations of Computer Science (FOCS), pp. 299–308. IEEE, Piscataway (2010)
76. Kumar, A., Sabharwal, Y., Sen, S.: A simple linear time $(1+\epsilon)$-approximation algorithm for k-means clustering in any dimensions. In: Proceedings of 45th Annual IEEE Symposium on Foundations of Computer Science, 2004, pp. 454–462. IEEE, Piscataway (2004)
77. Kumar, S., Mohri, M., Talwalkar, A.: Sampling techniques for the nystrom method. In: Artificial Intelligence and Statistics, pp. 304–311 (2009)
78. Kumar, S., Mohri, M., Talwalkar, A.: Sampling methods for the nyström method. J. Mach. Learn. Res. **13**, 981–1006 (2012)
79. Langberg, M., Schulman, L.J.: Universal ϵ-approximators for integrals. In: Proceedings of the Twenty-First Annual ACM-SIAM Symposium on Discrete Algorithms, pp. 598–607. SIAM, Philadelphia (2010)
80. Laurent, B., Massart, P.: Adaptive estimation of a quadratic functional by model selection. Ann. Stat. **28**, 1302–1338 (2000)
81. Le, Q., Sarlós, T., Smola, A.: Fastfood-approximating kernel expansions in loglinear time. In: Proceedings of the International Conference on Machine Learning, vol. 85 (2013)
82. Lee, H., Battle, A., Raina, R., Ng, A.Y.: Efficient sparse coding algorithms. In: Schölkopf, B., Platt, J.C., Hoffman, T. (eds.) Advances in Neural Information Processing Systems, vol. 19, pp. 801–808. MIT Press, Cambridge (2007)
83. Lee, J.R., Gharan, S.O., Trevisan, L.: Multiway spectral partitioning and higher-order cheeger inequalities. J. ACM **61**(6), 37 (2014)
84. Lei, J., Rinaldo, A.: Consistency of spectral clustering in stochastic block models. Ann. Stat. **43**(1), 215–237 (2015)
85. Li, M., Lian, X.-C., Kwok, J., Lu, B.-L.: Time and space efficient spectral clustering via column sampling. In: Proceedings of the 2011 IEEE Conference on Computer Vision and Pattern Recognition, pp. 2297–2304. IEEE Computer Society, Washington (2011)
86. Li, Q., Liu, W., Li, L., Wang, R.: Towards large scale spectral problems via diffusion process. In: 2017 International Conference on Digital Image Computing: Techniques and Applications (DICTA), pp. 1–7 (2017)
87. Lloyd, S.: Least squares quantization in PCM. IEEE Trans. Inf. Theory **28**(2), 129–137 (1982)
88. Lorenzo, P., Barbarossa, S., Banelli, P.: Chapter 9 - sampling and recovery of graph signals. In M. Djurić, P., Richard, C. (eds.), Cooperative and Graph Signal Processing, pp. 261–282. Academic Press, Cambridge (2018)
89. Loukas, A.: How close are the eigenvectors of the sample and actual covariance matrices? In: Proceedings of the 34th International Conference on Machine Learning. Proceedings of Machine Learning Research, International Convention Centre, Sydney, 6–11, vol. 70, pp. 2228–2237 (2017)
90. Loukas, A.: Graph reduction with spectral and cut guarantees. arXiv preprint arXiv:1808.10650 (2018)

91. Loukas, A., Vandergheynst, P. Spectrally approximating large graphs with smaller graphs. In: International Conference on Machine Learning (ICML) (2018)
92. Loukas, A., Simonetto, A., Leus, G.: Distributed autoregressive moving average graph filters. IEEE Signal Process. Lett. **22**(11), 1931–1935 (2015)
93. Makarychev, K., Makarychev, Y., Razenshteyn, I.: Performance of Johnson-Lindenstrauss transform for k-means and k-medians clustering. arXiv:1811.03195 [cs] (2018)
94. Malkov, Y.A., Yashunin, D.A.: Efficient and robust approximate nearest neighbor search using hierarchical navigable small world graphs. arXiv:1603.09320 [cs], p. 13 (2016)
95. Martin, L., Loukas, A., Vandergheynst, P.: Fast approximate spectral clustering for dynamic networks. In: Dy, J., Krause, A. (eds.) Proceedings of the 35th International Conference on Machine Learning, Stockholm, Sweden. Proceedings of Machine Learning Research, vol. 80, pp. 3423–3432. PMLR (2018)
96. Mohan, M., Monteleoni, C.: Beyond the nystrom approximation: speeding up spectral clustering using uniform sampling and weighted kernel k-means. In: Proceedings of the Twenty-Sixth International Joint Conference on Artificial Intelligence IJCAI (2017)
97. Mohan, M., Monteleoni, C.: Exploiting sparsity to improve the accuracy of nyström-based large-scale spectral clustering. In: 2017 International Joint Conference on Neural Networks (IJCNN) (2017)
98. Muja, M., Lowe, D.G.: Scalable nearest neighbor algorithms for high dimensional data. IEEE Trans. Pattern Anal. Mach. Intell. **36**, 2227–2240 (2014)
99. Munteanu, A., Schwiegelshohn, C.: Coresets-methods and history: a theoreticians design pattern for approximation and streaming algorithms. Künstl. Intell. **32**, 37–53 (2017)
100. Musco, C., Musco, C.: Recursive sampling for the nystrom method. In: Advances in Neural Information Processing Systems, pp. 3833–3845 (2017)
101. Newling, J., Fleuret, F.: Fast k-means with accurate bounds. In: International Conference on Machine Learning, pp. 936–944 (2016)
102. Ng, A., Jordan, M., Weiss, Y., et al.: On spectral clustering: analysis and an algorithm. Adv. Neural Inf. Process. Syst. **2**, 849–856 (2002)
103. Nister, D., Stewenius, H.: Scalable recognition with a vocabulary tree. In: 2006 IEEE Computer Society Conference on Computer Vision and Pattern Recognition (CVPR'06), vol. 2, pp. 2161–2168 (2006)
104. Ostrovsky, R., Rabani, Y., Schulman, L.J., Swamy, C.: The effectiveness of Lloyd-type methods for the k-means problem. In: 2006 47th Annual IEEE Symposium on Foundations of Computer Science (FOCS'06), pp. 165–176 (2006)
105. Paratte, J., Martin, L.: Fast eigenspace approximation using random signals. arXiv preprint arXiv:1611.00938 (2016)
106. Pena, R., Bresson, X., Vandergheynst, P.: Source localization on graphs via ℓ_1 recovery and spectral graph theory. In: 2016 IEEE 12th Image, Video, and Multidimensional Signal Processing Workshop (IVMSP), pp. 1–5 (2016)
107. Peng, R., Sun, H., Zanetti, L.: Partitioning well-clustered graphs: spectral clustering works! In: Conference on Learning Theory, pp. 1423–1455 (2015)
108. Philbin, J., Chum, O., Isard, M., Sivic, J., Zisserman, A.: Object retrieval with large vocabularies and fast spatial matching. In: 2007 IEEE Conference on Computer Vision and Pattern Recognition, pp. 1–8 (2007)
109. Puy, G., Tremblay, N., Gribonval, R., Vandergheynst, P.: Random sampling of bandlimited signals on graphs. Appl. Comput. Harmonic Anal. **44**(2), 446–475 (2018)
110. Rahimi, A., Recht, B.: Random features for large-scale kernel machines. In: Advances in Neural Information Processing Systems, pp. 1177–1184 (2008)
111. Ramasamy, D., Madhow, U.: Compressive spectral embedding: sidestepping the SVD. In: Cortes, C., Lawrence, N.D., Lee, D.D., Sugiyama, M., Garnett, R. (eds.) Advances in Neural Information Processing Systems, vol. 28, pp. 550–558. Curran Associates, New York (2015)

112. Rangapuram, S.S., Mudrakarta, P.K., Hein, M.: Tight continuous relaxation of the balanced k-cut problem. In: Ghahramani, Z., Welling, M., Cortes, C., Lawrence, N.D., Weinberger, K.Q. (eds.) Advances in Neural Information Processing Systems, vol. 27, pp. 3131–3139. Curran Associates, New York (2014)
113. Ros, F., Guillaume, S.: ProTraS: a probabilistic traversing sampling algorithm. Expert Syst. Appl. **105**, 65–76 (2018)
114. Saade, A., Krzakala, F., Zdeborová, L.: Spectral clustering of graphs with the Bethe Hessian. In: Ghahramani, Z., Welling, M., Cortes, C., Lawrence, N.D., Weinberger, K.Q. (eds.) Advances in Neural Information Processing Systems, vol. 27 pp. 406–414. Curran Associates, Inc. (2014)
115. Safro, I., Sanders, P., Schulz, C.: Advanced coarsening schemes for graph partitioning. J. Exp. Algorithmics **19**, 2 (2015)
116. Sakai, T., Imiya, A.: Fast spectral clustering with random projection and sampling. In: Perner, P. (ed) Machine Learning and Data Mining in Pattern Recognition. Lecture Notes in Computer Science, vol. 5632, pp. 372–384. Springer, Berlin (2009)
117. Sakiyama, A., Tanaka, Y., Tanaka, T., Ortega, T.: Eigendecomposition-free sampling set selection for graph signals. arXiv:1809.01827 [eess] (2018)
118. Sandryhaila, A., Moura, J.: Big data analysis with signal processing on graphs: representation and processing of massive data sets with irregular structure. IEEE Signal Process. Mag. **31**(5), 80–90 (2014)
119. Sculley, D.: Web-scale k-means clustering. In: Proceedings of the 19th International Conference on World Wide Web - WWW'10, Raleigh, p. 1177. ACM Press, New York (2010)
120. Sedley, D.: An introduction to Plato's theory of forms. R. Inst. Philos. Suppl. **78**, 3–22 (2016)
121. Shi, J., Malik, J.: Normalized cuts and image segmentation. IEEE Trans. Pattern Anal. Mach. Intell. **22**(8), 888–905 (2000)
122. Shuman, D.I., Vandergheynst, P., Frossard, P.: Chebyshev polynomial approximation for distributed signal processing. In: 2011 International Conference on Distributed Computing in Sensor Systems and Workshops (DCOSS), pp. 1–8. IEEE, Piscataway (2011)
123. Shuman, D.I., Narang, S.K., Frossard, P., Ortega, A., Vandergheynst, P.: The emerging field of signal processing on graphs: extending high-dimensional data analysis to networks and other irregular domains. IEEE Signal Process. Mag. **30**(3), 83–98 (2013)
124. Spielman, D.A., Srivastava, N.: Graph sparsification by effective resistances. SIAM J. Comput. **40**(6), 1913–1926 (2011)
125. Spielman, D.A., Teng, S.-H.: Spectral sparsification of graphs. SIAM J. Comput. **40**(4), 981–1025 (2011)
126. Stoer, M., Wagner, F.: A simple min-cut algorithm. J. ACM **44**(4), 585–591 (1997)
127. Sutherland, D.J., Schneider, J.: On the error of random Fourier features. arXiv:1506.02785 [cs, stat] (2015)
128. Tarsitano, A.: A computational study of several relocation methods for k-means algorithms. Pattern Recognit. **36**(12), 2955–2966 (2003)
129. Tremblay, N., Puy, G., Borgnat, P., Gribonval, R., Vandergheynst, P.: Accelerated spectral clustering using graph filtering of random signals. In: IEEE International Conference on Acoustics, Speech and Signal Processing (ICASSP) (2016)
130. Tremblay, N., Puy, G., Gribonval, R., Vandergheynst, P.: Compressive spectral clustering. In: 33rd International Conference on Machine Learning, New York (2016)
131. Tremblay, N., Amblard, P.O., Barthelmé, S.: Graph sampling with determinantal processes. In: 2017 25th European Signal Processing Conference (EUSIPCO), pp. 1674–1678 (2017)
132. Tremblay, N., Barthelme, S., Amblard, P.-O.: Optimized algorithms to sample determinantal point processes. arXiv:1802.08471 [cs, stat] (2018)
133. Tremblay, N., Gonccalves, P., Borgnat, P.: Chapter 11 - design of graph filters and filterbanks. In: Djurić, P.M., Richard, C. (eds.) Cooperative and Graph Signal Processing, pp. 299–324. Academic Press, Cambridge (2018)
134. Tsitsvero, M., Barbarossa, S., Di Lorenzo, P.: Signals on graphs: uncertainty principle and sampling. IEEE Trans. Signal Process. **64**(18), 4845–4860 (2016)

135. Ulrike von Luxburg, O.B., Belkin, M.: Consistency of spectral clustering. Ann. Stat. **36**(2), 555–586 (2008)
136. Vishnoi, N.K., et al.: Lx= b. Found. Trends Theor. Comput. Sci. **8**(1–2), 1–141 (2013)
137. Vladymyrov, M., Carreira-Perpinan, M.A.: Fast, accurate spectral clustering using locally linear landmarks. In: 2017 International Joint Conference on Neural Networks, pp. 3870–3879 (2017)
138. Von Luxburg, U.: A tutorial on spectral clustering. Stat. Comput. **17**(4), 395–416 (2007)
139. Wagner, D., Wagner, F.: Between min cut and graph bisection. In: Borzyszkowski, A.M., Soko\lowski, S. (eds.), Mathematical Foundations of Computer Science 1993, Berlin, Heidelberg, pp. 744–750. Springer, Berlin (1993)
140. Wang, Y., Feng, Z.: Towards scalable spectral clustering via spectrum-preserving sparsification. arXiv preprint arXiv:1710.04584 (2017)
141. Wang, L., Leckie, C., Ramamohanarao, K., Bezdek, J.: Approximate spectral clustering. In: Theeramunkong, T., Kijsirikul, B., Cercone, N., Ho, T.-B. (eds.), Advances in Knowledge Discovery and Data Mining. Lecture Notes in Computer Science, vol. 5476, pp. 134–146. Springer, Berlin (2009)
142. Wang, J., Wang, J., Ke, Q., Zeng, G., Li, S.: Fast approximate k-means via cluster closures. In: Multimedia Data Mining and Analytics, pp. 373–395. Springer, Berlin (2015)
143. Wei, Y.-C., Cheng, C.-K.: Towards efficient hierarchical designs by ratio cut partitioning. In: 1989 IEEE International Conference on Computer-Aided Design(ICCAD), pp. 298–301 (1989)
144. Wierzchoń, S.T., K\lopotek, M.A.: Spectral clustering. In: Modern Algorithms of Cluster Analysis, pp. 181–259. Springer, Cham (2018)
145. Woodruff, D.P., et al.: Sketching as a tool for numerical linear algebra. Found. Trends Theor. Comput. Sci. **10**(1–2), 1–157 (2014)
146. Wu, L., Chen, P.-Y., Yen, I. E.-H., Xu, F., Xia, Y., Aggarwal, C.: Scalable spectral clustering using random binning features. arXiv:1805.11048 [cs, stat] (2018)
147. Yan, D., Huang, D., Jordan, M.I.: Fast approximate spectral clustering. In: Proceedings of the 15th ACM SIGKDD International Conference on Knowledge Discovery and Data Mining, KDD'09, Paris, pp. 907–916. ACM, New York (2009)
148. Zhang, K., Tsang, I.W., Kwok, J.T.: Improved nyström low-rank approximation and error analysis. In: Proceedings of the 25th International Conference on Machine Learning, pp. 1232–1239. ACM, New York (2008)
149. Zhao, Z., Liu, H.: Spectral feature selection for supervised and unsupervised learning. In: Proceedings of the 24th International Conference on Machine Learning - ICML'07, Corvalis, pp. 1151–1157. ACM Press, New York (2007)

Chapter 6
Sampling Technique for Complex Data

A. Idarrou and H. Douzi

6.1 Context and Motivations

In the context of Big Data, the data is potentially unlimited in number. An explosion of complex data (text, video, audio, animation, 2D/3D graphics, etc.) due to advances in information and communication technology: Cloud computing technology reduces storage costs and significantly improves the capacity to process large amounts of data. However, all this data is useless if it cannot be used [37]. Most of this data is unstructured and comes from multiple sources in a variety of formats, most of which do not conform to a consistent data model. Yet, today, it is impossible to do without the automatic processing of increasingly voluminous and complex data.

With the emergence of complex data, which has affected most domains of activity, other issues are appearing and other fields of research are also emerging. In most domains, the data to be processed to extract knowledge from them is increasingly complex and voluminous [22]. In today's world, we are bring to manipulate complex data, in overabundance, too little or no structured and often dynamic and unlimited. Specifically, processing complex data in the context of Big Data is becoming increasingly unavoidable. Indeed, the volume of data increases by 40% each year, these data are at the service of business development of companies and organizations. More precisely, data has now become the raw material and the DNA of several application fields; this multiplies the need for methods and techniques allowing access to relevant information in a targeted and efficient way.

A. Idarrou (✉) · H. Douzi
Labo. Image et Reconnaissance de Formes – Systèmes Intelligents et Communicants (IRF-SIC), Ibn Zohr University, Agadir, Morocco
e-mail: a.idarrou@uiz.ac.ma; h.douzi@uiz.ac.ma

F. Ros, S. Guillaume (eds.), *Sampling Techniques for Supervised or Unsupervised Tasks*, Unsupervised and Semi-Supervised Learning,
https://doi.org/10.1007/978-3-030-29349-9_6

The advent of the Internet of Things (IoT), a booming developer field, has challenged conventional systems of data processing non-scalar. According to the Gartner Institute, it is planned that by 2020, more than 50 billion things will be connected. This is a real digital revolution that will certainly change our habits and more generally our way of life. The Internet of Things is a technique is imposed from day by day within organizations, entreprises and also in smart homes (connected Tv, connected refrigerator, etc.). This technique allows connected objects to exchange data with each other and to communicate certain information in real time. For example, in medical telemonitoring a large number of sensors connected to devices (electrocardiogram, blood pressure monitor, etc.) are necessary to monitor the condition of an elderly person, sick, or losing autonomy. The information fusion from different sensors (or other devices) allows crossing several parameters in order to guarantee as much as possible a resulting information that can be useful for remotely monitoring the state of these persons. More precisely, connected objects are able to collect data, interact with the environment, and send data in real time, so that it can be explored, analyzed, etc. In general, complex data comes from different sources such as sensors, drones, social networks, GPS, connected things, etc. These data are heterogeneous, of different size and nature. Accordingly, the problem of integration and/or data fusion arises. These data offer opportunities for success and open up new opportunities for businesses. Using scientific techniques, data analysis allows to detect trends and, more generally, strategic information that is often hidden in a large amount of data. However, the problem of non-scalar data processing involves several levels: volumetry, heterogeneity, multi-source character, size often unknown, etc. The analysis of these data is now at the center of the concerns of actors in most domains of activity whose predictions and perspectives are decisive.

With the appearance of data streams, another type of proliferating complex data that has profoundly affected most areas. In these areas, data is growing faster than our ability to explore and analyze it in real time. For example, in many sectors (road and air traffic monitoring, detection of anomalies in industrial chains, etc.), large data streams need to be processed in real time to provide near-immediate results. To face up this phenomenon, several works have applied a treatment on the fly on these data.

Exploiting in real time the full potential of the data available today becomes a major challenge. This often leads to the adaptation and/or extension of existing conventional methods and techniques, or to develop of new ones, especially for analysis processes and more generally for the exploitation of mass data. The classical approaches that have shown their effectiveness in processing scalar data need to be rethought. In fact, the emergence of new forms of representation of more or less complex data has shown the limits of these approaches. These must be adapted to respond to the particularities of complex and dynamic data in order to effectively support the expectations of the algorithms that one wish to apply on these data. The major challenge today is therefore how to structure, explore, analyze, and process massive data and variants over time? What methods and techniques are used to fully and effectively exploit the value of such data?

In this section, we focus more on works that have used sampling as a solution for processing complex data. We are approaching a series of works in order to provide a global vision of the issues and challenges related to the phenomenon of massive data.

6.2 Sampling on Complex Data

6.2.1 Preambule

Using the entirety of a large mass of complex data, before analyzing and querying it, becomes today a challange. More than this, respond to real-time queries becomes very costly in terms of storage, computation and latency imposed by the exploration of this data generally heterogeneous and dynamic. However, real-time query processing has become an increased need in many domains (security, medicine, social networks, military, aeronautics, marketing, road traffic, etc.). In a number of activity sectors, the real challenge in the Big Data era is to be able to process and analyze mass data from different sources in both structured and unstructured form in real time to extract value [40]. Several works have used sampling as a solution to overcome this problem. To perform an exploratory analysis on massive data, one approach is to approximate the result of a query [8, 33, 35].

The objective of sampling theory is to optimize the volume of data to be analyzed according to various criteria. In [17], in order to respond to the computational pressures and computing power of data streams, approximate approaches have been proposed, the most common of which is sampling: a simple and effective method. Sampling is a technique derived from statistics that can provide information on a large population from a representative sample of it.

Sampling can be used in large data situations to reduce the volume to be processed. In [36], one possible approach to processing large-scale data is to take a sample and explore it, as approximate responses are acceptable for many data mining applications. The aim is to work on a sample which is sufficiently representative of a population to extrapolate the result and transpose it on this population. In these situations, the theory of inferential statistics is used to infer laws. It is therefore a question of inducing from the sample to data set in its entirety.

In the continuous data concept, sampling is the procedure for selecting a representative sample of these data [10] and [48]. However, using a sample to process very large data can lead to unreliable results [34]. In this context, several works have extended classical methods of analysis to facing up to the Big Data phenomenon [16, 45, 57], etc. More generally, the evolution of large complex data is generating a greater need for extending existing data analysis methods to deal with mega-data.

Overall, two types of sampling are distinguished in the literature: probabilistic (or random) sampling and non-probability sampling.

6.2.2 Non-Probability Sampling

Non-probability sampling is also called empirical sampling. It is characterized by the fact that the probabilities of inclusion of the sample elements are unknown. In this type of sampling, subjective means are used to determine the extent to which a sample is representative of the entire population.

6.2.3 Probability Sampling

The probability sampling technique is based on the principle of randomization, according to which each element of the study population has a fair chance of being included in the sample. Generally, probability sampling techniques are expensive and time consuming than empirical techniques.

In the following sections, we describe the probability sampling methods applied in the context of complex data.

6.2.3.1 Simple Random Sampling Technique

Simple random sampling is the most simple and understandable basic sampling selection process. All individuals in the study population have the same chance of being included in the sample extracted from this population. In [18], with simple random sampling a sample is randomly selected so that each element of the population has the same probability of being selected. This sampling technique is appropriate when the population is relatively homogeneous (Fig. 6.1).

An advantage of simple random sampling is the ease of selecting the samples. It is also considered a fair way to select a sample from a given data set because each

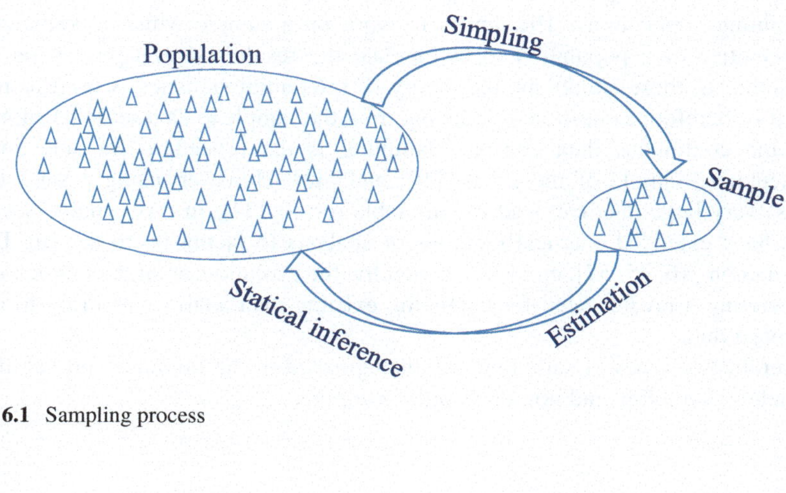

Fig. 6.1 Sampling process

element has the same probability of being selected. Consequently, a simple random sample provides an unbiased representation of the study population.

6.2.3.2 Stratified Sampling Technique

This stratified sampling technique consists in dividing a heterogeneous population into relatively homogeneous subgroups called strata. The choice of strata consists in obtaining homogeneity within each stratum. In practice, this technique is complex. On the other hand, it ensures that each subgroup of the population studied is represented in a sample extracted from this population (Fig. 6.2). Therefore, this method provides a representative sample.

6.2.3.3 Cluster Sampling Technique

The technique consists of subdividing a relatively homogeneous population into clusters. Then, a number of clusters are randomly selected to represent the whole population. The sample obtained is composed of all the units included within the selected clusters. This sampling technique is extremely less costly as it requires a minimum effort to construct a sample. The cluster sampling technique gives good results when the clusters are similar and the individuals in the same cluster are different.

As illustrated by the figure (Fig. 6.3), the population is divided into clusters that are all essentially the same as each other. Each cluster is essentially representative of the population as a whole (Table 6.1).

To end up this discussion, in cluster sampling, all the units of the randomly selected clusters make up a sample, while in the stratified sampling technique; the sample is created from the random selection of elements from all strata. Specifically, in the cluster sampling technique, a cluster is the sampling unit instead of a single element of the population.

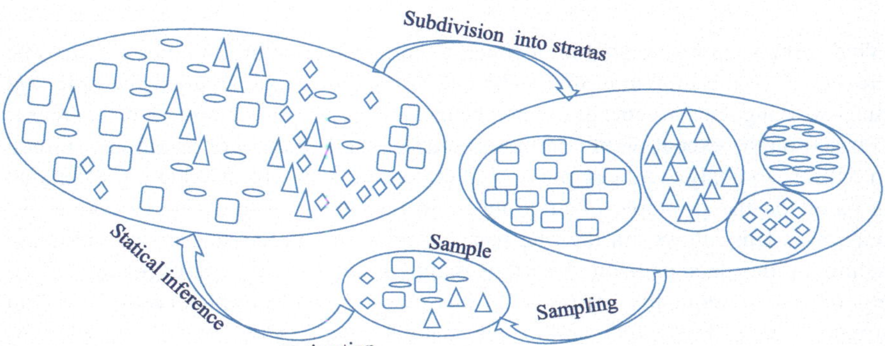

Fig. 6.2 Stratified sampling process

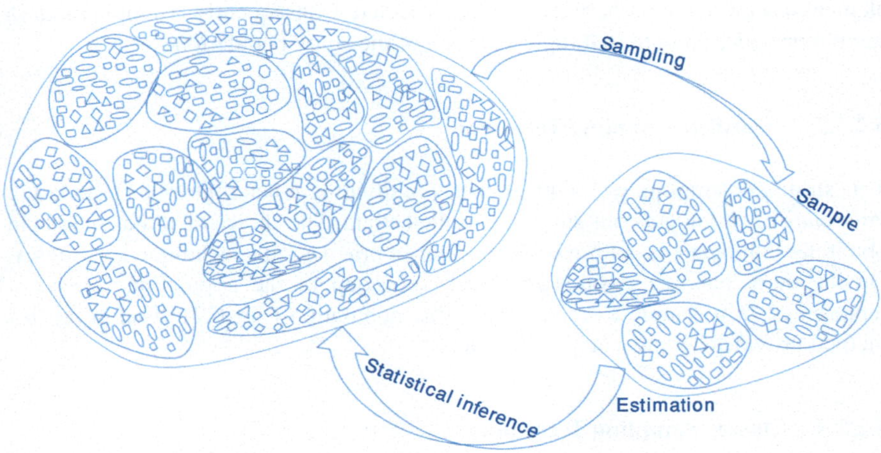

Fig. 6.3 Cluster sampling process

Table 6.1 Represents a comparative study between cluster sampling and stratified sampling

	Stratified sampling	Cluster sampling
Objective	To increase precision and representation	To reduce cost and improve efficiency
Sample	Randomly selected individuals are taken from all the strata	All the individuals are taken from randomly selected clusters
Selection	Individually	Collectively
Homogeneity	Within group	Between groups
Heterogeneity	Between groups	Within group

In the next section, we present an overview of the window sampling technique.

6.2.3.4 Window Sampling Technique

Windowing sampling is another technique that can reduce the amount of data to be processed. In the literature, this technique is an appropriate solution for sampling time-evolving data streams of unknown size. Indeed, the processing of data streams is based on the execution of continuous queries on data that are potentially infinite in nature. To analyze a large mass of data continuously, the windowing technique is based on the principle of dividing these data into a succession of successive windows. This allows, on the one hand, to adapt the processing to the unlimited nature of the data and, on the other hand, to improve the performance of the algorithms subsequently executed. In fact, a window is a subinterval of the data stream, where the desired sample size is much smaller than that of this window.

In [27], windowing techniques are used to adapt to the unlimited nature of the data. In the literature, several works have used the technique of sliding window, the technique of fixed window, etc. A fixed window is a subset of the set of continuous data, whose bounds are fixed such as a window between 8 am and 6 pm from Monday to Friday of each week, a window containing the data of a month, etc. On the other hand, the bounds of a sliding window change over time. The windowing technique can provide adequate solutions to maintain a random sample on a sliding window [9] (Fig. 6.4).

A window can be either physical when it is defined by means of time intervals (days, hours, minutes, seconds, milliseconds, etc.), or logical when defined in terms of number of elements. The windowing technique allows an approximate response to a query on the streaming data and allows the query to be evaluated not on the entire history of that data, but rather on the windows of the most recent data. Applied to this type of data, the windowing principle allows continuous queries to provide responses over limited time periods. In [47], a continuous query on data streams at a very high rate can only provide an answer over a necessarily limited period of time. Consequently, queries for relatively long period data and historical data cannot be processed. Several data streams management systems (DMFS) have been developed as examples: Stream [7], Aurora [1], etc. These systems allow to express continuous requests that evaluate progressively on a data stream or on parts of it. In a number of areas where data is generated continuously and ephemeral (data circulating on social networks, for example), processing is based on the time or order of continuous data. The size of the window should be large enough that window sampling is relatively representative and meaningful.

Windowing techniques allow for temporal limits on data streams. One of the advantages of this technique is that it gives exact results to a continuous query on the current window data. However, we cannot have the data history beyond the window. But in the context of data flows, it is interesting to consider the history of

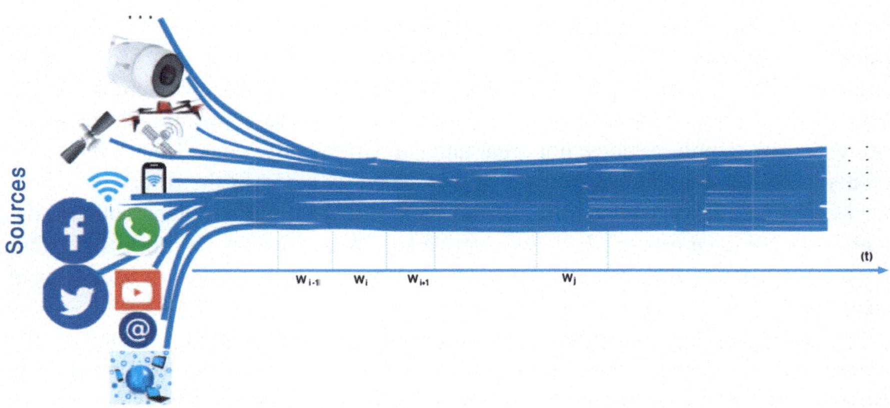

Fig. 6.4 Example of sliding window

the data in the sampling process. In [26], one solution is to keep a summary of the stream history to answer queries. Several data flow summary techniques have been developed. Reference [14] proposed a generic and optimized method of constructing summary from distributed data streams. The approach of [14] is based on the general summary as defined by [17] while taking into account all data streams.

In the literature, many works have addressed data streams summary structures [14, 17, 24, 26, 28, 31, 32], etc. The author of [17] has defined a summary of data stream according to two views: (1) a functional view of the summary where a summary aims to gather some of the information contained in the flow over a given period of time, to be able to solve a particular problem during this period. (2) a more general view where a summary aims to keep compact information on a data stream in its entirety temporal, without considering any problem a priori. The authors of [47] have defined a generalist summary as a data structure updated as the elements of a data stream arrive, and allowing for a posteriori response and approximately to queries about this data stream.

In the next section, we present an overview of the reservoir sampling technique.

6.2.3.5 Reservoir Sampling Technique

This sampling technique is particularly adapted to summarize large volumes of data [52]. Advanced sampling to sample data stream is reservoir sampling [56] and [12]. In fact, the reservoir algorithm was designed by Alan Waterman (called algorithm R) and then classified by [56]. The reservoir sampling requires a reservoir whose size corresponds to the size of the resulting sample. This technique has been successfully applied in many domains of activity: social networks (Twitter) in real time [13], text flow [12], etc. However, defining the size of the reservoir is difficult [41]. In this context and to address the dynamics of the data arriving with a high frequency, the authors of [4] proposed a dynamic adjustment of the size of the reservoir while the sampling is still in progress.

In [41], the authors proposed a sampling algorithm based on the reservoir sampling technique. This sampling algorithm whose objective is to process data streams without limits and which does not require to know the length of the flow of these data or the size of the sample. However, the conventional technique of reservoir sampling does not guarantee a sufficiently convinced number of (statistically) tuples of each portion of data to be included in the reservoir, which may call into question the quality of the sample [5]. To palliate this problem, the authors of [6] have proposed a new adaptive stratified reservoir sampling algorithm on several parts of the data to be processed. However, this algorithm uses a fixed size tank.

Several sampling algorithms are based on reservoir sampling. One of the advantages of reservoir sampling technique is that it reduces the load on the fly. However, this technique is limited when the data expires in a sliding window [15]. In fact, elements that are no longer part of the current window become invalid, and if they belong to the sample, they must be replaced. However, this may have an impact on the representativeness of the sample.

In fact, reservoir sampling is a family of randomized algorithms for randomly choosing a sample, composed of k elements, from a data set of n elements ($k < n$), where n is either a very large or infinite. Typically n is large enough that the entire data set doesn't fit into main memory.

The basic idea of the reservoir sampling algorithm is as follows:

(a) Create an array Reservoir[1..k]
(b) Reservoir[] Initialization: copy the first k *DataStream* elements into the Reservoir[]
(c) Then update Reservoir[] by randomly selecting elements from $e_{k+1}, e_{k+2}, \ldots, e_n$

Algorithm 1 Algorithm for reservoir sampling

Let DataStream=$\{e_1, e_2, e_3, \ldots, e_n\}$ be a Data stream of n elements (with n very large or infinite)
Reservoir[] is the output array
for all $i \in [1, k]$) **do**
    Reservoir[i]=e_i //Initialization
end for
// Iterate from the (k+1)th element to nth element
for all $i \in [i, n]$) **do**
    j = rand() % (i+1)
    if $(0 < j \leq k)$ **then**
        Reservoir[j]=e_i //If the randomly picked index is smaller than k, then replace
    end if
end for

6.2.4 Conclusion

To end up this discussion, the sampling technique is widely used to analyze large data. Results are then obtained with some acceptable approximation, and confidence limits can be given to the degree of error introduced by the sampling process [35]. However, extracting a sample sufficiently representative from a very large of complex and dynamic data is not always trivial. Several questions can be asked concerning the selection of this sample:

- How to construct a representative sample to ensure meaningful results?
- What is the most appropriate technique for building a sample?
- What is the optimal size of a sample?
- What is the validity period of a sample?
- How to maintain the representativeness of a sample of a large mass of complex and evolving data over time?

- How to apprehend the heterogeneity of the complex data of both dynamic and multiple sources?
- How precise should the results be?
- What are the constraints to be taken into account of the fact the unlimited size of massive data?
- and so forth

In the next section, we discuss the problem of constructing a sample of large complex data.

6.2.5 Construction of Sample

Practically, it is not always possible to analyze a mass of large size data or of infinite size in its entirety. Sampling is one of the most widely used techniques to overcome this problem. This consists of using a reduced representation (sample) but producing the same analytical results or approximate results. In the sampling process, the construction of a significant sample is a crucial step and the representativeness of the results depends on the quality of this sample. In the main, the method of building a sample must be based on an approach and a strategy. In [43], the construction of a sample is a strategic and evolving operation. In [26], the choice of a particular summary method depends on the nature of the data to be processed and the problem to be solved. Indeed, the results of an analysis based on a bad sample are of no interest. On the contrary, the interpretation of these results can even have negative and even catastrophic consequences.

In order to construct a justified and meaningful sample, it is necessary to define in a global way the strategy, the needs but also the constraints to be taken into account. Next, several questions need to be asked about the approach and techniques to be used to construct a representative sample of the population to be studied. Sampling a large mass of dynamic and evolving data over time is a very complicated task. Adding to this is the representativeness of a sample, which is a problem underlying the sampling of complex data flows.

The heterogeneity of complex data from multiple sources is one of the major difficulties in sampling this type of data. In fact, the management of inconsistency and uncertainty of data is a central step in the process of sampling massive data from heterogeneous sources. To overcome this problem of heterogeneity, several works have used the integration and data fusion to consolidate these data and prepare them to facilitate their analyses and queries.

6.2.6 Data Integration and Fusion

Data integration is an implicit problem in the problem of sampling data from heterogeneous sources. It is a process in which heterogeneous data, from multiple sources, is retrieved and combined as a unified form and structure. The underlying idea is to merge these data in order to reduce both the heterogeneity of these data and the uncertainty of the resulting data. Data fusion refers to a set of methods and techniques for aggregating data from heterogeneous sources. It aims to obtain the most complete information possible through complementarity and redundancy of information from multiple sources. In [3], data fusion consists of combining several observations of an environment or phenomenon to produce a more complete description (Fig. 6.5).

In [11], information fusion combines information from several sources to improve decision-making. Indeed, the large number of heterogeneous distributed sources and the amount of complex data generated by them have made it necessary to integrate and merge these data in order to obtain a unified and relatively complete data. In a number of application areas where information sharing is unavoidable, there is an increased need to consolidate and federate massive data from multiple and heterogeneous sources at different levels: format, type, dynamic structure, etc. Nonetheless, it is difficult to tackle the problem of the heterogeneity of complex data in its entirety. In [53], the heterogeneity of data sources is generally classified into two categories: (1) heterogeneity of structures and (2) heterogeneity of the content of these data. To overcome this problem, several works have used integration as a process to reduce this heterogeneity. In fact, integration is a solution allowing unified access to data from multiple heterogeneous sources. This technique provides unified access to different sources of information while masking the heterogeneity,

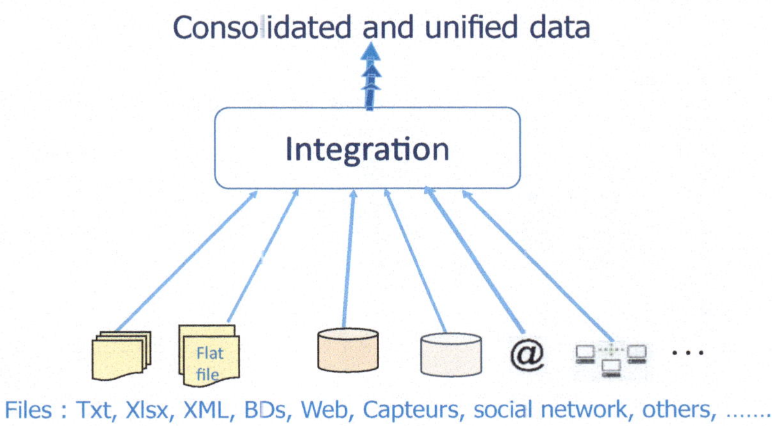

Fig. 6.5 Data integration process

autonomy, and distribution of these sources. In the literature, there are two main approaches to data integration.

6.2.6.1 Data Warehouse Approach

The data warehouse approach consists of a materialized integration where data from multiple heterogeneous sources are stored (after extraction, transformation, and loading) into a data warehouse.In the context of the data warehouse approach, the integration is a process allowing to construct a huge database called data warehouse storing data from multiple heterogeneous sources [53]. More precisely, data integration involves combining data from several multiple distributed sources, which are stored using various technologies. It is a process in which heterogeneous data is extracted and combined as an unified form and structure (Fig. 6.6).

The integration system following a data warehouse approach is composed of three layers:

(a) Sources
It contains data that is generally heterogeneous in terms of format, structure, semantics, etc.

(b) Data warehouse
It is a large database where data is extracted from sources, cleaned, and transformed according to the warehouse schema and finally stored in the warehouse.

(c) Data mart
It is a lightened extract from the data warehouse whose purpose is to focus on a subject, a theme or a business in order to respond the needs of a user group. In other words, data mart is a small data warehouse that satisfies the needs of a reduced set of users.

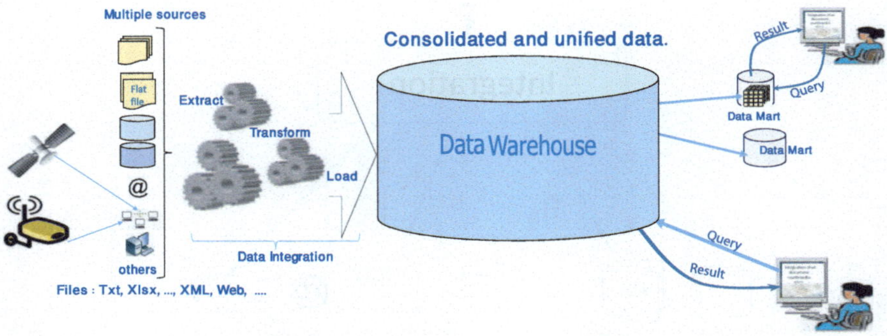

Fig. 6.6 Data integration process: Datawarehousing

Concretely, data warehouses have demonstrated their ability to integrate a large quantity of data and their efficiency in querying this data. Indeed, the analysis of stored data is based on data organization techniques in the form of multidimensional databases. More generally, the data warehouse is recognized like the decisional system core; it integrates and stores the data in order to become easily accessible to the processes of decisional analyses [49]. Several works have used the data warehouse approach to consolidate heterogeneous data [19, 37, 44, 46], etc. In [2], the companies and the organizations resort to data warehousing systems in order to provide decision makers a comprehensive view and synthetic information circulating in their organizations. However, in the context of Big Data, it becomes very difficult to manage massive data using data warehouses based on relational databases. In fact, the emergence of new forms of complex data representation may call into question the existing conventional systems. More precisely, conventional data warehouses are inappropriate for complex data stream. In order to tackle this phenomenon, integration systems must be adapted. The advent of NoSql (Not-Only SQL) databases is an alternative solution for building multidimensional data warehouses that can support large amounts of data available today. In Fact, NoSQL databases are an interesting way to build multidimensional data warehouses that can support large amounts of data [23].

In order to benefit from the advantages of cloud computing technology in terms of computing power, response time and cost reduction, other work [51] and [39] proposed to implement data warehouses in cloud computing.

6.2.6.2 Mediator Approach

This technique consists of a virtual integration where the data remain in the sources. The interrogation of these data goes through a global scheme. The user feels he is questioning a homogeneous system. Queries are expressed on the global schema, then decomposed into subqueries (a sequence of queries) on the sources. The results of the sources are combined to form the final result (Fig. 6.7).

In [30], mediators are intermediary modules in large-scale information systems that link multiple sources of information to applications. In other words, a mediator is a query interface between the user and data sources. It is a mediation system that constitutes an architectural solution for transparent access to distributed heterogeneous sources [42]. The first mediator architecture has been proposed in [29], it is represented by three layers:

(a) Mediator

at the mediator level, the global schema provides a unified vocabulary for the expression of user requests. For the description of the contents of the sources is represented by a set of abstract views on the sources. In other words, the mediator allows to locate the data and solve the schematic conflicts.

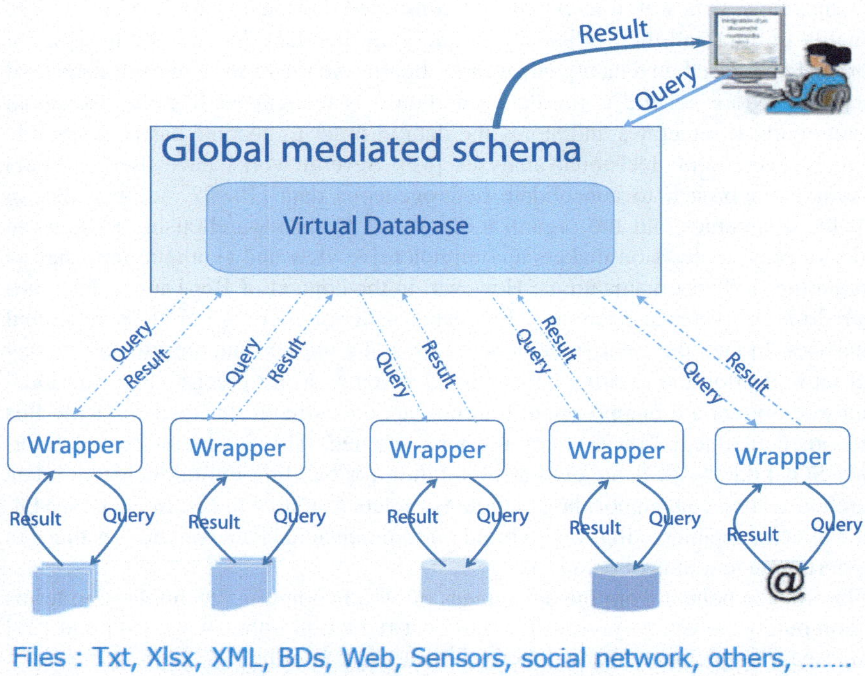

Fig. 6.7 Data integration process: mediator architecture

(b) Wrappers
at the wrappers level,the requests are expressed in terms of the vocabulary of the global schema (mediator) in queries expressed in the source language. More precisely, the wrapper allows resolving data conflicts by presenting them in the mediation model
(c) Sources
The sources are queried directly through the wrappers and mediator.

In reality, the mediator does not store the source data, but it has abstract views of the data stored in the sources. To query data at the mediator level, mappings must be defined between global schema relationships and local schema (sources) relationships. Thus, wrappers transform responses to queries into responses that conform to the global mediator schema. More precisely, the effective querying of the sources is done through wrappers which translate the requests expressed by the views into the query language specific to each of the sources (Table 6.2).

To benefit from each of the approaches, several works like [20] and [38] have proposed to combine both approaches: data warehouse and mediator. In principle, data warehouse tends to be faster since the view is materialized before the query is run. The user interacts with the warehouse for a direct query of the warehouse

Table 6.2 Represents a comparative study between data warehousing approach and mediator approach

Parameters	Warehousing approach	Mediator approach
Storage of the interrogated data	Data warehouse	Data sources
Data query	DataWarehouse or Data mart	Data sources
Response time	Fast	Relatively long
Updating data query	Complex and often long	Not necessary

data or through the data mart. Mediator, on the other hand, provides up-to-date information because data sources are directly queried and the view is not evaluated before execution, so the response time is slower.

In the next section, we discuss the problem of representativeness of a sample in the context of complex data.

6.2.7 A Sample Representativeness

A sample of a population is representative when it has the same characteristics as the entire population. In other words, a representative sample can produce representative results that can be extrapolated to the study population. And the quality of the results of a statistical analysis of the data, strongly depends on the representativity of the sample of the population studied. The representativeness of a sample is fundamental to the data analysis process. Indeed, the results of the analysis of a sample will not be justified when the sample is not representative of the entire population to be studied. However, maintaining a representative sample in the context of massive, dynamic, and complex data is a difficult task. In principle, the problem of sample representativeness is underlying the sampling process.

In [50], a sample is representative of a large population when from there one can describe this population. More precisely, a sample is representative when it comes from a representative print [50]. Clearly, if the sample is not randomly selected, it may not be representative of the population to be studied. In statistics, a random sample is a sample drawn at random from a population in which all elements have the same chance of being included in this sample. For [55], the definition of a representative sample differs depending on whether the sample is probabilistic or not. In [54], a sample is never representative in itself, it is representative in relation to certain variables. In [21], a sample constructed using the quota method is representative. Following the quota method, to construct a sample, we get to have a representative sample [25].

With the emergence of massive data that is continuously evolving over time, sampling techniques face a challenge. One of the major challenges is maintaining the representativeness of the sample, due of the temporality of the data, their evolution, and their heterogeneity. Indeed, maintaining the representativeness of a sample is a problem that raises several questions about the order of dynamic data, the time history of these data, the time it takes to update the sample, etc. As a result, there is a growing need for methods and techniques capable of processing massive and dynamic data of very large size in a more efficient way in time and precision.

6.3 Review and Synthesis

Nowadays, it is impossible to do without the automatic processing of more and more voluminous and complex data. The mass data processing problem is accentuated with the advent of data, dynamic, scalable and from different sources with an infinite size that exceeds our ability to process them in real time. The analysis of complex large data in real time is a hot topic of growing interest in several key areas of activity (security, military, medicine, legal, etc.) whose prediction and / or decision-making are extremely important. In concrete terms, the processing of massive data is a rapidly developing promising domain. It is a central focus of the concerns of the actors of the most part of the activity domains. In a number of application areas (eg airport security), it is necessary to analyze a large mass of complex and continuous data in order to exploit the results in real time. Therefore, there is a need for both tools and techniques from statistics, mathematics and automatic learning. In this context, serveral startups has used data analysis as one of the core elements of their strategy.

More generally, the real challenge in the era of Big Data is to be able to explore and analyze complex and dynamic data from different sources in real time, in both structured and unstructured form. In essence, these data are heterogeneous in terms of type, nature, format, size, structure, etc. To overcome the problem of heterogeneity, several works have used data fusion to reconcile and consolidate data from heterogeneous in order to obtain a resulting and coherent data. In fact, data fusion is the process of aggregating data from multiple sources to reduce uncertainty about the resulting data.

In this chapter, we have discussed some sampling techniques on complex data. For example, windowing techniques and reservoir sampling techniques. One of the advantages of windowing techniques is that they provide exact answers to a continuous query on the current a data window. yet, we cannot have the data history beyond the current window. However, in continuous data analysis, we cannot do without data (of other windows) that does not belong to the current window. The windowing technique must be enriched with models taking into account the history of the data but also to adapt to the data unlimited nature. This will allow to consider the specificities of the data: dynamic and evolutive as over time, etc. Confronting the results of successive windows may be of great importance to discuss critical issues.

Reservoir sampling has been applied successfully in many areas, such as real-time social network feeds, for example. However, one of the limitations of this technique is the problem of defining the size of a reservoir. In this context, a few works have proposed a dynamic adjustment of the reservoir size while sampling is still in progress. This allows facing up the dynamics of continuous data. Other works have proposed sampling algorithms based on the reservoir sampling technique, which allow for continuous processing of data of unknown size.

In the end, the processing of complex data raises many problems related to the exploitation of these data in real time. Not all conventional sampling tools and techniques are adapted to the processing of complex data streams. Sampling algorithms must evolve to taking into account the specifics of complex data streams of unknown size. It is therefore necessary to have mathematical, statistical, and machine learning tools capable of processing data on a large scale. Several works have extended the methods of analysis to face up of Big Data phenomenon. On the other hand, few works have raised the problem of data semantic heterogeneity. Indeed, the autonomy of data sources can also lead to semantic heterogeneity. In general, each source can use a specific vocabulary in its context, which can lead to misinterpretation of the data generated by it. More specifically, semantic heterogeneity can lead to ambiguities when data are interpreted by data analysis systems. Not considering semantic heterogeneity in a data analysis system could have a negative impact on the results. For example, the fact that a considerable number of messages that circulate on social networks and cannot be interpreted correctly is a handicap for social network analysis. The integration of semantic annotation into complex data sampling process, using dedicated ontologies, could provide a solution to this type of problem and therefore increase the relevance and quality of the results.

Concretely, the design of sampling-based algorithms that can produce approximate responses close to the exact answer is an active research domain. This is an opportunity for both industrialist and academic researchers. Currently, the problem of sampling complex data streams has led to numerous investigations by researchers in order to propose solutions taking into account the specificities of this data type. Existing solutions today are not yet perfect, other ideas need should be explored.

References

1. Abadi, D.J., Carney, D., Cetintemel, U., Cherniack, M., Convey, C., Lee, S., Stonebraker, M., Tatbul, N., Zdonik, S.: Aurora: a new model and architecture for data stream management. VLDB J. **12**(2), 120–139 (2003)
2. Abdelhédi, F.: Conception assistée d'entrepôts de données et de documents XML pour l'analyse OLAP. Thèse de Doctorat de l'université de Toulouse (2014)
3. Andre, C.: Approche crédibilise pour la fusion multi capteurs décentralisée. Thèse de Doctorat Université Paris Sud (2013)
4. Al-Kateb, M., Lee, B.S., Wang, X.S.: Adaptive-size reservoir sampling over data streams. In: Proceedings of the International Conference on Scientific and Statistical Databases Management, pp. 22–34, Washington, DC (2007)
5. Al-Kateb, M., Lee, B.S.: Stratified reservoir sampling over heterogeneous data streams. In: International Conference on Scientific and Statistical Database Management, pp. 621–639 (2010)
6. Al-Kateb, M., Lee, B.S.: Adaptive stratified reservoir sampling over heterogeneous data streams. Inf. Syst. **39**, 199–21 (2014)
7. Arasu, A., Babu, S., Widom, J.: Cql: a language for continuous queries over streams and relations. Lecture Notes in Computer Science, pp. 1–19. Springer, Berlin (2004)
8. Avnur, R., Hellerstein, J.M.: Eddies: continuously adaptive query processing. SIGMOD Rec., **29**(2), 261–272 (2000)

9. Babcock, B., Babu, S., Datar, M., Motwani, R., Widom, J.: Models and issues in data stream systems. In: PODS '02: Proceedings of the Twenty-First ACM SIGMOD-SIGACTSIGART Symposium on Principles of Database Systems, pp. 1–16. ACM, New York, NY (2002)
10. Bibiduh, L., Tirthapura, S.: Stream sampling. Encyclopedia of Database Systems, pp. 2838–2842. Springer, Berlin (2009)
11. Bloch, I.: Fusion d'informations en traitement du signal et des images. In: IC2, Trait IC2. Hermes Science, Paris, France (2003)
12. Bonin, R., Marcacini, R.M., Rezende, S.O.: Unsupervised instance selection from text streams. J. Inf. Data Manag. **5**(1), 114–123 (2014)
13. Chen, C., Yin, H., Yao, J., Cui, B.: Terec: a temporal recommender system over tweet stream. Proc. VLDB Endow. **6**(12), 1254–1257 (2013)
14. Chiky R.: Résumé de flux de données distribués, thèse de Doctorat de l'Ecole Nationale Supérieure des Télécommunications, France (2009)
15. Chiky, R., Hébrail, G.: Echantillonnage optimisé de données temporelles distribuées pour l'alimentation des entrepôts de données. In 3èmes Journées Francophones sur les Entrepôts de Données et l'Analyse en Ligne (EDA 2007), Poitiers, vol. B-3 of RNTI, pp. 51–66. Cépaduès, Toulouse (2007)
16. Chu, C., Kim, S.K., Lin, Y.-A., Yu, Y., Bradski, G., Ng, A.Y., Olukotun, K.: Map-reduce for machine learning on multicore. Adv. Neural Inf. Proces. Syst. **19**, 281 (2007)
17. Csernel B.: Résumé généraliste de flux de données, Rapport de thèse. ENST, Paris (2008)
18. Cochran, W.G.: Sampling Techniques. John Wiley et Sons, Hoboken (2007).
19. Dejmal, K.: De la modélisation à l'exploitation des documents à structures multiples. Thèse de Doctorat de L'Université de Toulouse, France (2010)
20. de Moura, A.M.C., Victorino, M., Tanaka, A.: Combining mediator and data warehouse technologies for developing environmental decision support systems. Lecture Notes in Computer Science book series (LNCS), vol. 2478 (2012)
21. Dutarte P.: L'induction statistique au lycée (ed : Didier) (2005)
22. ECG.: Fouille de Données Complexes dans un processus d'extraction de connaissances. Extraction et Gestion des Connaissances, Lille (2006)
23. El Malki, M.: Modélisation NoSQL des entrepôts de données multidimensionnelles massives. Thèse de Doctorat de l'Université de Toulouse (2016)
24. Feraud, R., Clérot, F., Gouzien, P.: Sampling the join of streams. In: IFCS'2009 International Conference on International Federation of Classification Societies, Dresden (2009)
25. Fourquet J.: Emission C dans l'air du 17 Février (2011)
26. Gabsi, N.: Extension et interrogation des résumés de flux de données. Thèse de Doctorat de l'Ecole Nationale Supérieure des Télécommunications, France (2011)
27. Gabsi, N., Clérot, F., Hébrail, G.: Résumé hybride de flux de données par échantillonnage et classification automatique. Résumé hybride de flux de données par échantillonnage et classification automatique. In: EGC '09: Conference Internationale Francophone sur l'Extraction et la Gestion des Conaissances, Strasbourg, France (2009)
28. Gemulla, R., Lehner, W.: Sampling time-based sliding windows in bounded space. In: SIGMOD '08: Proceedings of the 2008 ACM SIGMOD International Conference on Management of Data, pp. 379–392. ACM, New York, NY (2008)
29. Gio, W.: Mediators in the architecture of future information systems. Computers **25**(3), 38–49 (1992)
30. Gio, W.: Mediators, Concepts and Practice To appear in Studies Information Reuse and Integration in Academia and Industry. Springer-Verlag, Wien (2012)
31. Guha, S., Harb, B.: Wavelet synopsis for data streams: minimizing non-euclidean error. In: KDD 2005: Proceedings of the Eleventh ACM SIGKDD International Conference on Knowledge Discovery in Data Mining, pp. 88–97. ACM, New York, NY (2005)
32. Guha, S., Koudas, N., Shim, K.: Data-streams and histograms. In: STOC '01 : Proceedings of the Thirty-Third Annual ACM Symposium on Theory of Computing, pp. 471–475. ACM, New York, NY (2001)

33. Haas, P.J., Hellerstein, J.M.: Ripple joins for online aggregation. SIGMOD Rec. **28**(2), 287–298 (1999)
34. Haddad R.:Apprentissage supervisé de données symboliques et l'adaptation aux données massives et distribuées. Thèse de Doctorat de l'Université de Paris dauphine, France (2016)
35. Hellerstein, J., Haas, P., Wang, H.: Online aggregation. In: Proceedings of the 1997 ACM SIGMOD International Conference on Management of Data, pp. 171–182 (1997)
36. Hu, W., Zhang, B.: Study of sampling techniques and algorithms in data stream environments, 2012. In: 9th International Conference on Fuzzy Systems and Knowledge Discovery. Sichuan, China (2012)
37. Idarrou, A.: Entreposage de documents multimédias : comparaison de structures. Thèse de Doctorat de L'Université de Toulouse, France (2013)
38. John, S.: Feeding a data warehouse with data coming from web services. A mediation approach for the DaWeS prototype. Other [cs.OH]. Université Blaise Pascal - Clermont-Ferrand II (2014)
39. Karkouda, K., Harbi, N., Darmont, J.: Gérald Gavin Confidentialité et disponibilté des données entreposées dans les nuages. In: 12ième Conférence Internationale Francophone sur l'Extraction et la Gestion de Connaissance (EGC 2. 2012)
40. Karoui, M., Devauchelle, G., Dudezert, A.: Big Data : Mise en perspective et enjeux pour les entreprises, N Spécial "Big Data". Revue Ingénierie des Systèmes d'Information **19**(3), 73–92 Hermès (2014)
41. Kepe, T.R., de Almeida, E.C., Cerqueus, T.: KSample: dynamic sampling over unbounded data streams. J. Inf. Data Manag. **6**(1), 32–47 (2015)
42. Kostadinov, D., Peralta, V., Soukane, A., Xue, X.: Intégration de données hétérogènes basée sur la qualité. Dans les actes des 23e Congrès INFORSID, Grenoble, France, pp. 471–486 (2005)
43. LeCompte, M.D., Preissle, J.: Ethnography and Qualitative Design in Educational Research. Academic Press, San Diego (1993)
44. Loic, M., Loudcher, S., Boussaïd, O.: Analyse en ligne d'objets complexes avec l'analyse factorielle. 10èmes Journées Francophones d'Extraction et de Gestion des Connaissances (EGC 2010), pp. 381–386. Tunisie (2010)
45. Malek, M., Kadima, H.: Searching frequent itemsets by clustering data: towards a parallel approach using mapreduce. In: Web Information Systems Engineering–WISE 2011 and 2012 Workshops, pp. 251–258. Springer, Berlin (2013)
46. Mbarki, M.: Gestion de l'hétérogénéité documentaire: le cas d'un entrepôt de documents multimédia. Thèse de Doctorat de L'Université de Toulouse, France (2008)
47. Midas, C.D.A., Inria, A.: Résumé généraliste de flux de données. EGC: Extraction et Gestion des Connaissances, Jan 2010, pp.255–260. Hammamet, Tunisie (2010)
48. Muthukrishnan, S.: Data Streams: Algorithms and Applications, vol. 1. Now Publishes Inc. (2005)
49. Olivier, T.: Modélisation et manipulation des systèmes OLAP : de l'intégration des documents 'a l'usager, Interface homme-machine [cs.HC]. Université Paul Sabatier - Toulouse III (2009)
50. Padieu, R.: Qu'est-ce que la représentativité? In: Economie et statistique, vol. 56, pp. 65–66 (1974)
51. Petre, R.-S.: Data mining in cloud computing. Database Syst. J. **3**(3), 67–71 (2012)
52. Raissi, C., Poncelet, P.: Echantillonnage pour l'extraction de motifs séquentiels : des bases de données statiques aux flots de données. EGC, pp. 145–156 (2008)
53. Sais, F.: Intégration Sémantique de Données guidée par une Ontologie. Thèse de Doctorat de l'Université de Paris Sud (2007)
54. Sautory, O.: Journée d'études sur la représentativité, ENS Paris (2010)
55. Vaillant, J.: Initiation à la théorie de l'échantillonnage, Web (2005)
56. Vitter, J.S.: "Random sampling with a reservoir". ACM Trans. Math. Softw. **11**(1), 37–57 (1985)
57. Zhao, W., Ma, H., He, Q.: Parallel k-means clustering based on mapreduce. In: IEEE International Conference on Cloud Computing, pp. 674–679. Springer, Berlin (2009)

Chapter 7
Boosting the Exploration of Huge Dynamic Graphs

F. Javier Calle, Dolores Cuadra, Jesica Rivero, and Pedro Isasi

7.1 Introduction

Almost three centuries ago, Euler published a paper in which he stated the problem of the bridges of Königsberg with which the study of graphs began, yet they were not named like that until a century and a half later. Since then, graphs have been a versatile mathematical structure widely used for modeling diverse types of real-world problems. Its many areas of application, ranging from natural sciences (chemistry, physics, or biology) to linguistics and social sciences, entail that any progress made in graph processing multiplies its utility since it will be reflected in improvements over many solutions to as many problems. Consequently, there is a great interest in developing algorithms for handling graphs efficiently.

Specifically, the classic "path search" problem over graphs or networks has been widely analyzed and discussed for many years. Advances in computer science and technology entailed also the evolution of both the supporting technological resources and the requirements, forcing research on the techniques applied for its resolution.

On the one hand, huge networks comprising thousands (even millions) of nodes and arcs have to be observed. The size and complexity of such giant graphs brings to mind the possibility of fragmenting and preprocessing the graph. However, our current frequent needs regarding these problems include observing dynamic graphs, which reduces our possibilities of preprocessing.

Indeed, these enormous networks are often subjected to frequent changes, affecting either the cost or even the existence of both nodes and arcs. For example, the task of routing packages through a communication network can be affected by a

F. J. Calle (✉) · D. Cuadra · J. Rivero · P. Isasi
Computer Science Department, Carlos III University of Madrid, Madrid, Spain
e-mail: fcalle@inf.uc3m.es

F. Ros, S. Guillaume (eds.), *Sampling Techniques for Supervised or Unsupervised Tasks*, Unsupervised and Semi-Supervised Learning,
https://doi.org/10.1007/978-3-030-29349-9_7

server going down, and by cost rocketing of a certain connection between two nodes due to network saturation. Even the very definition of the service, searching a path connecting two nodes through the network, can vary by changing one of the nodes to be connected. Because of such dynamicity, in certain domains the validity of the solution could be expiring in time, so the agility in both providing and applying the solution is a strong requirement.

Classic approaches sought best path (shortest or cheapest), but the risk of validity expiration leads to seek fast algorithms, privileging response time over cost. Besides, in many domains, the time taken for covering a path is lower than the time required to find a better solution. Anyhow, it will be valuable to observe algorithms able of providing a fast solution and of refining it when more time is available. Moreover, the solution can be refined while currently being applied. A particular case is when the path taken loses validity (because of dynamicity). Being able of revising that path for quickly finding a similar valid solution, instead of restarting the path search, will be most appreciated. This is a basic feature of ant colony optimization (ACO) algorithms [1, 2], which have been largely applied on dynamic (yet not quite big) graphs. These methods are found on the repetition of successful behaviors through several essays performed by a set of colony members. After a few iterations, the solution found could be not as good as the optimum but probably good enough and obtained in less time. Besides, such approach ensures good adaption to changes, either for the service parameters and the network structure.

On the other hand, the path search algorithms have usually been focused as isolated processing, performed locally on individual devices. However, there could be found certain advantages in setting servers which use common resources for all clients. For example, through guiding services provided to itinerant elements it will be required finding a path connecting their position with their goal, and it also could be useful taking into account the position of the remaining elements, sometimes for making them come together in a point, or perhaps for avoiding saturated arcs. Actually, not only their position but also their trajectories or recent positions could be of use, so predictions on graph dynamicity can be attained, and also partial solutions can be provided by another client seeking a similar or close goal.

Therefore, a certain challenge is to adapt these algorithms to be supported by database technologies, so they can benefit from a storage of higher capacity, durability, and range. In particular, this work presents the joint of two technologies, databases and ACO algorithms, and explores their adaptation and evolution for solving the current form of the need: fast search of paths through vast dynamic graphs.

7.2 Background

Graphs are a versatile tool commonly used for representing and formalizing problems of diverse type and nature. Over the years, and more profusely with the involvement of computer science, there have been proposed and developed many algorithms for processing graphs with different aims. Specifically, there are several approaches for path finding in graphs, of different nature and strategy. Departing from a graph defined as set of nodes N and a set of edges E (links from n_i to n_j with a cost c_{ij}, where $n_i, n_j \in N$), path finding problem is defined as providing a sequence of nodes (linked by edges) connecting a start node α with a target node ω. Solutions are paths characterized by a cost (the sum of the costs of all edges in the path), and reducing that cost (finding better paths) is the usual goal.

As aforementioned, classic algorithms (e.g., Dijkstra, $A*$, etc.) are aimed to check the existence of a path between two nodes, and seek the best solution (minimum cost). These approaches are often based on the analysis of the entire graph, which is a burden as the size of the subjected graphs grew.

Let us take Dijkstra's, as a classic widespread solution. It is a greedy algorithm proposed in 1959 by Edsger Dijkstra [3]. It is supported by a set of resolved nodes with their preceding node S (initialized to $\varnothing$), and a set A containing accessible nodes with provisional cost and preceding node (initialized to all nodes directly connected to α by an edge, with provisional preceding node α and provisional cost the one indicated by that edge). Until A is empty, it will remove from A the node n_i with minimum provisional cost p_i, adding that node n_i (with its preceding node) to S, and analyzing every edge from n_i to n_j with cost c_{ij}: if n_j is not in A, or if the provisional cost of n_j in A (p_j) is greater than $p_i + c_{ij}$, then it will add/update the node n_j in A, with preceding node n_i and provisional cost $p_i + c_{ij}$. After at most $x-1$ iterations (where x is the cardinality of N), the set S will support the solution for every node reachable from α (the path from α to ω will be obtained by taking preceding nodes in S from node ω to node α).

The evolution of those algorithms sought reducing the processing time, by applying some tactic. Among those techniques some should be highlighted, such as preprocessing the graph, fragmenting the graph into several smaller subgraphs, and also pruning the exploration of the graph by applying some heuristic. The latter are techniques that sacrifice optimality and even accuracy for attaining better response times or eventually to attain an approximate solution when exact solutions are unreachable. They are based on certain prior knowledge, which makes them domain specific. Though there could be cases in which the efficiency drops or even a solution is never attained, such cases are rare, and efficiency is good on average.

Finally, when dynamicity hinders preprocessing and it is not possible to find a seemly good heuristic, the general metaheuristic algorithms appear as a solution. *Metaheuristic* are generic (domain independent) versatile techniques guiding the solution search for finding quasi-optimal solutions with some efficiency gain and often other advantages regarding classic algorithms, such as the ability of adapting their operation to different scenarios, or even adaptation to changes produced in running time. Yet they are mostly non-deterministic, and they could be showing eventual anomalies, such behavior is really rare, and their general working is usually advantageous.

Regarding solutions based on preprocessing, some approaches (Bast's [4], Delling's [5]) create a hierarchical structure in main memory for supporting fast exploration. An evolution of the latter [6] divides the graph into subgraphs and runs a preprocessing (whose result is also stored in main memory) to determine which edges should be observed for each subgraph. More recently [7], these authors summarize algorithms applied in route planning for transportation networks, taking into account the resources spent in preprocessing, the space and computing requirements, and of course the response time. Basically, the algorithms solve point-to-point shortest path and the start point is that data fits in RAM. They assume that while preprocessing may take a long time (e.g., hours), queries need to be fast.

On the other hand, there are also several proposals supported by secondary storage, thus avoiding the problems caused by hardware resources constraints. Chan [8, 9] preprocesses the graph and divides it into subgraphs to be stored in raw files. Some works apply database technology to manage the secondary storage [10], which will serve for keeping the graph, the results of the storage, and a cache of solutions to frequently requested services. For example, Sankaranarayanan and Samet [11, 12] use databases for organizing graphs into tree structures, as a result of their preprocessing, which will enable processing huge graphs in less time.

As a conclusion, it should be pointed out that almost all similar works are supported by a costly preprocessing (several days time) performed before starting up the service. Besides, most of those algorithm are focused on running the analysis on the whole graph, requiring to be completely restarted whenever any structural change is performed in the graph (dynamic graph conditions). As a good prospect for the future, it would be advantageous to aim the development of new algorithms to local preprocessing, thus avoid preprocessing the entire graph when any structural change occurs. The proposal presented in [13] applies two algorithms for reducing the original graph. This proposal solves the specific problem in a communication network, where the response time is not the most important parameter to be optimized.

A typical dynamic graph example is that of the social networks. Regarding path search algorithms in this domain, it should be mentioned one of the first relevant works presented by Kleinberg [14] who performs a decentralized search observing only local data. Some further works extending that techniques are found in [15, 16], improving performance and observing more information such as the population distribution. Honiden [17] proposes an algorithm for vast graphs supported by Voronoi duals (graphs enunciated by Mehlhorn [18] and Erwing [19] inspired in classic Delaunay triangulation for Voronoi diagrams, whose size is smaller than the original graph). The search is performed over the Voronoi duals by applying the algorithm of Dijkstra, and the result (a path in the dual graph, not in the original one) is used for directing a search on the original graph. Main drawback is that Voronoi duals are created through a costly preprocessing. Recently, some proposals are focused on applying heuristic algorithms to prove the small-world structure in networks. The main idea shown in [20] is the calculation of the shortest path between any two nodes of the network. In this approach, the network is huge but the execution time is not relevant. This work deals with dynamics changes simulating these changes in the Twitter network. In [21], the string network typology is adapted to a small-world structure. The algorithm has a good performance but the experimentation is carried out in networks with moderate size (20–100 nodes).

As a nature-inspired technique [22], these algorithms are based on massive number of simple participants working in parallel but asynchronously. They do not require global information (could be decentralized) and, though the results come from their joint action, their communication mechanisms are very plain.

Specifically, *ants* (individual agents of simple behavior seeking global achievement) mark successful behaviors using pheromone which can be noticed by other individuals to follow (or not) such conducts [1, 2]. Hence, a simple individual can tackle a complex task in a large environment because in fact it is only observing local conditions, and it is not supposed to solve the problem: actually, the colony as a whole will be providing the solution. This approach has been applied to find paths across graphs, network routing, the traveling salesman, problem, etc.

When finding paths across graphs, an individual in a node will select an arc among many by the calculation of their probability based on the pheromone each arc has. On the other hand, the ants leave a slight trace of pheromone in the covered arc when looking for food (the target), and a strong trace when going back home (the start point). Thus, after several waves of ants, a seemly good path can be found by following the pheromone trace. The pheromone will be dissipated in time, but better decisions will be reinforced with new pheromone.

This method is not going to assure optimal cost path, as other classic approaches as Dijkstra's algorithm do. It is even not going to improve the initial response time. But it shows other features that could involve interesting results, such as robustness (is not crucial for an individual to get lost or collapsed), distribution, etc. Besides, stored preprocessing (not initializing the pheromone across the network) provides a way to the target from diverse start points, which is a very interesting added value. Finally, and above all, it presents strong adaptation to dynamic changes: addition or deletion of nodes or arcs will quickly be overcome, as well as the problem recalculation of the path when deviations in its execution appear.

Its versatility has been proved in many domains and variations: vehicle routing problem [23, 24], calculating trajectories for unmanned vehicles [25], project management [26, 27], improvement of software quality prediction models [28], medical risk profile recognition [29], cybersecurity in the web [30, 31], or social networks [32], among many other fields.

Regarding its adaptation to new resources and requirements, ACO has been adapted to exploit parallel environments [33, 34], to reduce the problem complexity applying local search in time windows [35, 36]. Finally, regarding its performance on dynamic scenarios, ACO has proven to be effective in facing changes [37], though in small formalizations and showing response times still far from real-time systems requirements.

Summarizing, the challenge is approaching an algorithm on huge dynamic graphs. Dynamicity will lead to set local preprocessing (if any) and privilege agility: it should be able of providing a fast first solution and refining it as more time is available. It will be supported by database technologies, as a shared resource by many concurrent services, so some benefits can also be explored while also new open questions and challenges will be left for further work.

7.3 Method Description

The aim of this work is to develop an algorithm able of proposing a path over a big dynamic network in a short response time (privileging agility over quality of the path). As a start point, the ACO family algorithms will be focused, because of their quick adaptation to changes in the network. Moreover, their working consists in progressively refining already found solutions, which suits perfectly with our goals. However, they are not prepared to work on big networks (on the contrary, their stochastic nature could be a strong hindrance for working on them). On the other hand, they are also not prepared to work in several concurrent services, which can take advantage of the processing done by another service (of different definition). So, the algorithm needs to be evolved to meet these requirements. In this section, the method development will be explained, leaving its formalization for a later section.

7.3.1 Ant Colony Base Algorithm

The ant colony optimization (ACO) method [2, 38] is a nature-inspired method emulating the behavior of colonies of ants. It is based on the joint behavior of a set of individuals that use simple rules to produce complex results through the repetition of behaviors by many individuals. Those behaviors, initially of random nature, are led by other individuals' discoveries, communicated by means of pheromone marks.

Ants will leave a light trace of pheromone as they look for their target, lighter as they move away. To avoid ants going in cycles, they can be endowed with a *taboo list* containing already visited nodes, thus preventing that ant from revisiting them. When an ant reaches the target, it will get a strong load of pheromone that will be left as it returns back home following the same path. That is, the path followed by the successful ant will be marked with pheromone, decreasingly from the target to the start point. After a few ant waves, the emergent behavior could be taken as a good result.

The proposed algorithm works on a graph, as defined at the beginning of Sect. 7.2, and proceeds as follows:

1. Initialization: all the networks arcs are endowed with a fixed pheromone amount.
2. All ants are placed onto the starting point node.
3. Ants are activated. They will individually perform the next steps:
   (a) Choose edge to follow: it will be a random choice from all available edges departing from current node (minus those leading to nodes included in the taboo list), but the probability of choosing any given edge will be proportional to the amount of pheromone found on it.
   (b) Update the pheromone of the chosen link (local update): the amount added will be inversely proportional to the cost of the edge.

   (c) Update ant position, and include last visited node into the taboo list.
   (d) If the ant is in target node or if it is stuck (every available link is in taboo list) deactivate ant (end). Else, go back to step (a).

4. When every ant is deactivated, perform global and dissipation updates:

   (a) Global update: for each ant in destination node and for all the links of its path, the link pheromone is increased in some amount inversely proportional to the path cost.
   (b) Dissipation: all the links reduce their amount of pheromone in a given percentage.

5. If the convergence conditions are not reached, go to step 1.

A first implementation of this method [39] running on a database (Oracle™ 10 g under Solaris 10) showed bad results in comparison to Dijkstra's classic method: not only the latter provides optimum path while ACO does not, it also finishes faster (only one of each eight services finished faster with ACO). In this case, the network was big enough ($2 \cdot 10^5$ nodes) for the ants to get lost, while small enough for a brute-force search to be efficient. Anyhow, the results pointed out that the DB took much time for operating nodes and links (DB needed tuning for improved performance, and reducing the amount of operations was found advantageous), and that randomly wandering ants get lost easily when working on big networks. Several changes were added sequentially to attain better performance.

Finally, it also should be stated that the experiments were performed over a "small" network. In fact, what is targeted is to find paths between two nodes across hundred thousand nodes networks. If the network is small enough, it would be no problem for the DB system to store best path between two random nodes and quickly retrieve it when required. But such procedure is not, of course, feasible for huge networks. The same experiments performed over a bigger network displayed even worse results: Dijkstra's algorithm spent nearly a minute where ACO's barely reached convergence.

7.3.2 First Approach: Splitting the Ants

One of the pointed burdens of the proposal is the high amount of DB updating operations. Anytime an ant goes somewhere at least one update is performed. Thus reducing covered surface of the network when first solution is met (any ant reaches the target node) will reduce significantly that response time.

Through the basic approach, when first solution is found the ants should probably had covered a circle around the origin node of radium *d*, being *d* the distance between origin and target nodes, as depicted in Fig. 7.1a. However, placing half of the ant set in the target node (searching for the origin node) and the other half of the set in the origin node (vice versa), they will be covering increasingly bigger circles until both circles get in touch, as shown in Fig. 7.1b. Then, the ant contacting

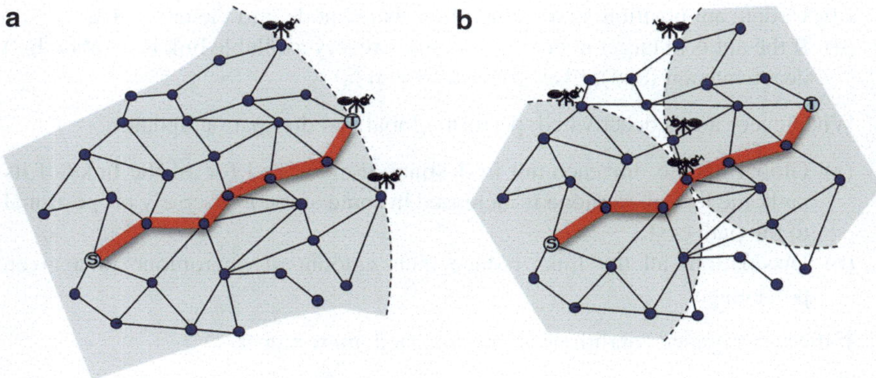

Fig. 7.1 (**a**) Surface covered by the ants through the basic approach. (**b**) Surface covered by the ants split into two sets

the other circle will probably follow the pheromone trace till it reaches its center, achieving a solution. At that moment, the surface covered by the whole set of ants will be slightly over two circles of radium *d*/2, which is nearly the half of the previous. Since smaller surface will probably contain fewer nodes, the updating operations should have been reduced significantly.

It must be noted that this variation, as explained, can be applied on non-directed graphs. In directed graphs arcs are unidirectional, and for applying the approach it is only required to have the ants from the source node following the arcs in direct way; and to have the ants from the target node taking the arcs in reverse way, that is, instead of observing arc departing from the node where they are, they have to consider all arcs arriving to that node.

Such scope leaves some space to improvement. For example, considering two different breeds of ants, with different pheromone traces, will enable an ant to privilege traces of the other breed, avoiding some probability of getting lost. Additionally, in certain domains the service definition can be partially repeated (for example, routing services frequently heading to a reduced set of target nodes) and splitting the ants into two colonies increases the probability of reusing the results of one of them for another service.

7.3.2.1 Improving the Effect of the "Taboo List"

Another interesting variation affects the taboo list. When any individual is *blocked* by taboo list it will be restarted. Promising individuals will probably follow pheromone trails to be soon found at a good state, but repeating those first steps takes time. What should actually be sought is not to prevent the ant going on loops (because if the ant finally succeeds, those eventual loops in the final path can be easily removed). Instead, what should be attained is preventing the ant to be taking the same choice in the same situation. That is, the ant is allowed going to an already visited node, but it will be forced going to an unexplored node in the next step. Thus, the ant can cross its own path for exploring the other side of the graph.

With this approach, the ants could go too frequently to already visited nodes or go zigzagging around its own past steps. Thus, the new taboo list approach is generalized to be based on a new parameter β. With this method (β-*taboo*) any ant will be allowed going anywhere (including already visited nodes) only if its β previous steps (at least) were performed to non-visited nodes. The variation explained at the beginning of this section should be consequently named *1-taboo*.

7.3.3 Second Approach: The Smell of Food

Initializing the network (removing all pheromone traces) is required for running a new service, because deposited pheromone was intended for a specific service (defined by start_node and target_node) and it will muddle the ants of the new colony set for a new different service. It is quite improbable to be repeating exactly the same service, and even to be receiving a similar service request for which all pheromone traces from former service are useful. However, while origin nodes are very diverse and nearly random for routing services, the target node of most services is usually included in a small set of *common targets*.

Therefore, it would be of interest to keep the pheromone associated with "commonly targeted nodes." For achieving this, it is proposed to observe two sorts of traces: on the one hand, pheromone traces left by ants while solving specific services; on the other hand, persistent traces indicating closeness to a target point. Following a nature-inspired idea, the new traces will be named "food essence" (FE), and they are left by another type of agents (β-ants) set on the target node (*food* node).

The β-ant traces not only should not be eliminated from a service to the sequent one, but also be initialized in a preprocessing: since the target point is known before the service request, the β-ant task could be done in advance. It should be noted that this FE traces will present as many classes as privileged target nodes (many, in general), and that structural changes due to dynamicity would entail updating them. Consequently, it is necessary to restrain in some way its dissemination in order to keep the volume of data update operations in a reasonable amount. It will be considered an influence area around the food node, whose size will be small regarding the graph but subject to specific analysis in each application case (part of the method parameterization). Anyhow, as a local preprocessing, its cost will not be high (and taken off-line, while the system is idle).

Definition 1 Food essence of node i in node j is a numerical value representing the distance between those nodes, and noted $O(i,j)$.

Thus, when any ant is seeking the target node i, in fact it will be searching for any node j with $O(i,j) > 0$. Then, following the increasing trace of FE_i, it will be easy for it to reach the target node.

This focus raises two issues: first, the risk of losing the adaptation to structural changes in dynamic networks, which characterizes original ACO algorithm; second, multiplying the pheromone data should not be a problem for a DB system suited to manage massive storing, but losing control of the β-ant traces could imply storing beyond technical possibilities: storing $O(i,j)$ $\forall_{i,j}$ means almost as many data as the number of nodes squared. It should be taken into account that the algorithm is aimed to huge networks, and such line could lead to an impracticable implementation.

Regarding first problem, the combination of ACO and DB technologies will provide the solution. Proper solution from ACO algorithms is found in the pheromone dissipation and later refreshment processes. However, dissipation is suitable for weakening probabilities but not for a trace that is going to be followed even if it is too weak. Thus, regarding FE traces, the traces removal will be performed through another process, the *evaporation*, consisting in eliminating FE traces when required.

At first sight, any structural change along the network will unleash the review of the $O(i,j)$ $\forall_{i,j}$ assignments. However, in this proposal the events involving such process are much more restricted than that. In fact, the evaporation and refreshment processes must be applied on a FE_i (food essence of node *i*) individually whenever:

- Any node *j* with $O(i,j) > 0$ is deleted or updated.
- Any link regarding a node *k* with $FE_{i,k} > 0$ is deleted, updated, or inserted.

If the proportion of FE_i across the network is low, which should be intended for huge networks, then the probability of FE_i evaporation and refreshment will be low as well. Last but not least, the DB provides the suitable mechanisms for *triggering* these actions where those events over the FE tables occur, which means high efficiency and low implementation effort for such operations.

Coming to the second problem, a very high volume of data in the FE tables, DB mechanisms (such as distribution) are hardly enough to alleviate the problem and do not shape a solid solution. Hence, a limit has to be placed to the FE_i propagation $\forall$ target node *i*, fixed through some parameters defined as follows:

Definition 2 The *amount of food essence* of node *i* is the initial charge of essence for any β-ant departing from node *i*, and noted $O(i,i) = m$. It is not obligatory to consider a constant *m* value $\forall$ food node *i* (frequently used or critical food nodes could have higher values).

Definition 3 The *odor loss* (k) fixes the proportion of food essence to propagate from one node with such odor to directly linked nodes. The *food essence* of that neighbor will be calculated as the food essence of its neighbor minus the product of odor loss and the cost of the link between both nodes.

Definition 4 The *food essence threshold* (μ) is the minimum quantity of FE of any kind for a β-ant to store it (lower values will not be stored).

At this point, the $O(i,j)$ calculation can be described as the maximum of all its candidate values. Calculating the food essence of node *i* in node *j* involves observing one candidate value for each link in which node *j* takes part.

These parameters should be valued seeking equilibrium between the algorithm efficiency and the DB performance. High $O(i,i)$ and k, and low μ, will increase the proportion of nodes marked with FE_i, thus easing α-ants to find a path to node i. However, high amount of nodes marked with FE will lead to DB low performance (depending also on the DBMS capabilities and the *Hw* resources).

Seeking the balance, what should be aimed is to mark some small percentage of the nodes with each FE. On the other hand, once a node marked with FE is reached, only the nodes arranging the best path to the target point are necessary, and none of their surroundings. In addition, that path is available when finishing a guiding service (after several rounds of ants).

Summarizing, what is being added to the method is that the final result of every guiding service (the whole path provided as final solution) will be marked with its FE, in the same way the β-ants do but ignoring the food essence threshold (μ). Hence, yet the FE marks reach very light values, they are enough to guide the α-ants. And much more network surface is covered with that FE by marking only a small percentage of the total nodes.

Graphically, the resulting schema could be described as a more or less small circle of food essence around each target node at the initial state. As the system is in use, there will appear several radiuses departing from it, helping ants set for further services to reach that food node. Taking a visual metaphor of food essence in the graph, it could be seen as the main roads departing and surrounding a big city, or the arteries in the circulatory system. Moving away from the center, where the radiuses are distant, there will also appear strings covering the distance between radiuses (as the minor roads reaching the main ones). The schema is depicted in Fig. 7.2 (left), along with the roads comparison (right). Thus, after the system is in use for some time, when any ant starts walking searching for a target node it will be easier for it to reach a node marked with its FE. In addition, the focus preserves its dynamic adaptation to structural changes performed over the network: whenever a node marked with FE is affected by any change, all the FE is evaporated. Not affected nodes will soon recover their FE, and the affected ones will fall into disuse.

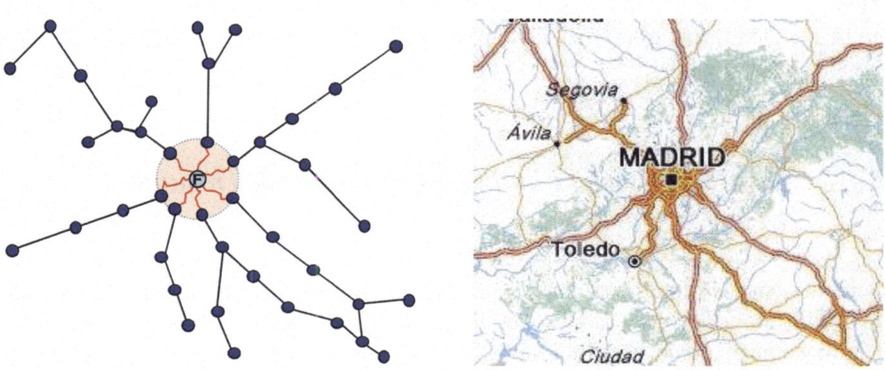

Fig. 7.2 A target node, its FE circle, and FE radiuses, as main roads around a city

Unfortunately, this new feature keeps on producing too much FE traces at medium term: almost every new service finishes leaving some new trace. Given that DB is crucial all along the system, and its performance should be assured, it is necessary to add a mechanism for controlling the FE tables' growth. A simple mechanism consists in fixing a boundary for those tables' volume. Hence, a new concept is introduced:

Definition 5 The *global volume of food essence threshold* (υ) is the maximum quantity of FE data to be stored (given the DBMS and Hw features).

Finally, a new process will be introduced, to eliminate the FE surplus. The *periodic evaporation* will be a process checking if the υ limit has been exceeded. In case, it will also calculate the amount of data to be deleted, and will choose and delete the lighter FE data from the tables. Such procedure is very simple to implement and run over a DB, by means of DB jobs, and keeps the privileges of nodes with higher $O(i,i)$ that will have longer radiuses and more strings. However, it also produces some degeneration at long term. Indeed, after producing lots of paths, evaporation could shorten even main radiuses and these will trend to disappear in benefit of the FE circle size, which will probably be higher than fixed by μ. Current solution consists in simply performing periodic (long term) evaporation of every FE trace over μ. It could be proposed some advanced FE marks including their usage frequency, so evaporation processes can be based not only in the food essence strength, but also in their usage, thus respecting main radiuses and eliminate less useful traces.

The approach, named SoSACO v1 (sense of smell ant colony optimization), was implemented and evaluated [39]. First evaluation observed only services targeting a food node (from any random start point, find a path reaching a given food node) were observed. In these conditions, and after preliminary experimentation for parameterizing the method in order to adapt it to the specific processing capabilities and resources of the evaluation environment, both path quality (cost) and response time were significantly improved. However, the strong limitations of the method (target node is always a predefined privileged node) weaken its applicability. Although it is true that in many domains there is a set of frequently requested nodes (to be reached from anywhere), it is aimed to attain an algorithm capable of solving any given service [40].

7.3.4 Third Approach: Finding a General Path Through a Food Node

General services should be defined as a path request from a random node to another one, with no concern to food nodes. Formerly, with the first approach (see Sect. 7.3.2), we have already focused that sort of services through a colony split into two

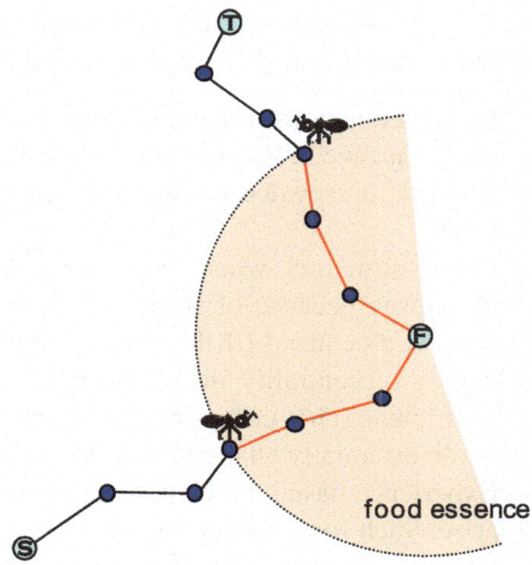

Fig. 7.3 Two breeds of ants quickly finding a path through the same FE_i

breeds, one for each node defining the service (start node and end node). Whenever any ant reached the pheromone of the other, it had found a path.

The basis of this new approach is the same but adding one food node and extending the success condition: whenever both breeds of ants reach traces of the food node, a solution is found. Certainly, finding traces of the food node involves that the ant, by following the incrementally stronger trail of food, has found a path from its origin to the food node. Since both breeds attain such paths, by chaining them they have jointly found a complete solution. If more time is available, that path will be marked with pheromone and ants will be able of refining it (for improving its quality). This idea is graphically depicted in Fig. 7.3.

Having a single food node in the entire (huge) graph could end up in many services taking too long for finding its food traces. Because of this, the method observes the existence of several food nodes, and the success condition is reformulated to be "when both breeds of ants reach traces of the same food node." The individuals (ants) defined for classic ACO algorithm are of simple processing and are not communicating with other individuals (except from the pheromone trails). For this variation, little communication is required: there will be a small *blackboard* (held by the DB, thus accessible for all ants) on which each breed of ants will be registering which food traces has found till far. Therefore, success condition can be quickly checked by any ant in the moment it finds a new trail. When the ant finds FE_i, it will check whether it is a new type of FE for its breed; in case it is, then it will check if the other breed has it; otherwise, it will add to the blackboard and continue; if the other breed has it, a solution is found.

The proliferation of FE_i could entail losing efficiency, so it is important to keep the balance between the processing capabilities and resources of the server and the parameterization of the food nodes (number of food nodes i, initial essence, essence loss $\Delta_{j,k}$, essence threshold μ, etc.).

Once parameterized, there is still the need of choosing appropriate nodes to be serving as food nodes. There could be found different criteria for this aim:

- Domain dependent: in many cases, it may be advantageous to choose nodes with a significance in the domain (landmarks): a famous or popular person in a social network, a crowded circus in the map of the town, or a privileged device in a LAN.
- Graph structure: when there is no domain knowledge, the food node would be chosen because of structural reasons: a node with outstanding connectivity degree; a centroid (if the network topology is metrizable); or a node with high degree of centrality in the graph [40, 41], that is, nodes with high clustering coefficient. This criterion can be applied on either static or dynamic scenarios while eventually changing the food node on the latter [42].
- Experience based: by inspecting the network usage some good candidates could arise. Such food node selection can be implemented to be taken and periodically updated automatically. For example, nodes with high transit frequency can be good meeting points, and good candidates to be chosen as food nodes.

In the next section, the explained method will be formalized, providing more details on formulas and agents behavior.

7.4 Method Formalization

As stated in the previous section, ACO method consists in creating a colony (of a certain number of individual) positioned in the start node, and let each individual wander through the network, following a probabilistic formula (Eq. (7.1), probability of going from node i to node j, where $i, j \in N$ and j is adjacent to i). This probability is proportional to the amount of pheromone at the link τ_{ij} multiplied by a heuristic value η_{ij}, and inversely proportional to the summation of that product ($\tau_{ix} \cdot \eta_{ix}$) for the rest of the candidate nodes.

Equation (7.1): Next node selection probability:

$$p(i, j) = \frac{[\tau_{ij}]^{e_1} \cdot [\eta_{ij}]^{e_2}}{\sum_{x \in \mathrm{Adj}(i)} [\tau_{ix}]^{e_1} \cdot [\eta_{ix}]^{e_2}}, \text{ where } i, j \in N \text{ and } j \in \mathrm{Adj}(i) \quad (7.1)$$

The amount of pheromone found at each node varies in time, due to three types of events: (a) anytime an ant uses a link (local update); (b) anytime a complete path is attained (global update); and (c) periodically all pheromone trails are decremented. Local update consists in increasing the pheromone deposited just after choosing a link, by adding a value inversely proportional to the cost assigned to the link (w_{ij}) and proportional to a constant value ($k_1/1000$) as shown in Eq. (7.2). Through global update, the pheromone is increased by a constant value (k_2) divided by the global cost of the complete path found, as formalized in Eq. (7.3). Finally, pheromone trails should be eventually removed causing unsuccessful attempts to be forgotten.

Specifically, just before any global update, the *evaporation* will be applied to all nodes in the graph, as stated in Eq. (7.4). These processes introduce some constant values (k_1, k_2, and the evaporation rate ρ) as parameters of the method that should be suited to each specific scenario through preliminary experimentation.

Eq. (7.2): Local update:

$$\tau_{ij}(t) = \tau_{ij}(t-1) + \frac{k1}{1000 \cdot w_{ij}} \tag{7.2}$$

Eq. (7.3): Global update:

$$\tau_{ij}(t) = \tau_{ij}(t-1) + \frac{k2}{\text{pathLength}} \tag{7.3}$$

Eq. (7.4): Evaporation:

$$\tau_{ij}(t) = (1-\rho) \cdot \tau_{ij}(t-1) \tag{7.4}$$

7.4.1 The Food Nodes

Given a food node $i \in N$, its trail (odor) at another node $j \in N$ is notated as $O(i,j)$. The maximum odor value is limited by parameter m, and is restricted to the food nodes themselves, that is, $O(i,i) = m \ \forall i \in N$. Regarding the minimum odor value, any trail below that value μ will not be recorded during the preprocessing, thus shaping a reasonable sized smell-area $S(i)$ around the food node i, and preventing the network to be saturated with the odor trails.

The preprocessing is supported by an auxiliary data structure, the bag B of nodes to be examined at any given iteration. The odor initialization algorithm starts with $B(0) \equiv \{i\}$, being i the focused food node. The preprocessing algorithm is as follows:

```
step 0   n:=0; B(n) ≡ {i}; /* initialization; first bag only includes the food node */
step 1   Bag for further iteration B(n+1) is empty (set to ∅);
step 2   For each node j ∈ B(n), let be R the set of nodes x∈N adjacent to j
             For each node x∈R,
                 calculate odor i coming from node j as O_j(i,x) = O(i,j) − k · w_jx
                 If new odor is greater than the currently stored O_j(i,x) > O(i,x)
                    and greater or equal than the odor threshold O_j(i,x) ≥ μ
                      then update odor O(i,x) := O_j(i,x) and include current node
                  x into B(n+1)
             end loop;
         end loop;
step 3   If the bag for further iteration is not empty B(n+1) ≠ ∅,
            then start a new iteration n := n+1, and go back to step 1;
         in other case (bag B(n+1) is empty), END.
```

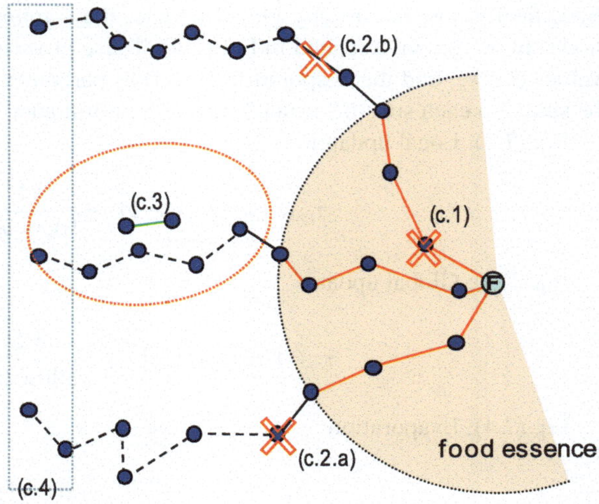

Fig. 7.4 Four cases of food essence evaporation

where k is a constant value between 0 and 1 (*odor loss*) and w_{jx} is the cost of the link between nodes j and x. Once this process ends (for each food node) it can be identified a non-empty subset of nodes $S_i \subseteq N$, namely the *odor coverage of node i*, where nodes $j \in S_i$ meet $O(i,j) \geq \mu$.

On the other hand, if some structural change is performed in the graph (dynamicity) the odor trail(s) found at the elements changed must be updated. Reducing the coverage area of the odor trail reduces the set of nodes with that benefit, but also enhances processing time and maintenance time (in case of dynamical graphs). The (food essence) evaporation processes will be applied, as shown in Fig. 7.4: full evaporation (for changes inside the *odor coverage* of the affected node), path evaporation (for individual element deletion), and link insertion. A fourth case will be observed for fixing periodic evaporation procedure.

(c.1) *Full evaporation*: when a node n_j is deleted, or a link involving a node n_j is deleted or inserted, having $O(i,j) \geq \mu$, the whole trace coming from it should be removed. That is, $O(i,x) := 0$, $\forall\, n_x, \exists\, w_{j,x} > 0 \wedge O(i,j) \geq O(i,x)$. If that node's food trail is removed, the same evaporation process will be applied to it, removing smaller food trails from its neighbors, and so on.

(c.2) *Path evaporation*: when the change affects a node belonging to a path, all food traces from that point on will be removed (the rest of the path, with lower odor than the removed node). Actually, the method prepared for the former case (c.1) suits perfectly with this other.

(c.3) *Arc insertion*: elements can also be added. Since odor is propagated through arcs, the case of interest is adding a new arc, between nodes a and b. In such case, the lower trail of food essence of the two connected nodes $min(O(i,a), O(i,b))$ should be reviewed (recalculated as the maximum between its current trail $O(i,a)$ and the general trail calculation $O(i,a) - (k \cdot w_{a,b})$). In case the trail is changed, it must be propagated to all its neighbors with a trail lower than that node's.

(c.4) *Periodic evaporation*: this evaporation process is a periodic one, to ensure data growth does not go beyond DB server possibilities. It introduces a new parameter measuring the amount of FE data (υ) that can be managed without efficiency loss. It will be fixed taking into account the server features (OS, DBMS, Hw, etc.) or empirically, by testing its response time with different loads of data and searching for an equilibrium (the more data, the better the algorithm works but worse response time of the server).

7.4.2 The SoSACO Algorithm

In the previous section, the ACO methods were adapted to the specific conditions and challenges that this new algorithm pursues, and thus some basis and techniques supporting the SoSACO algorithm were introduced in a sometimes intuitive and always non-formal way. This section tackles its formalization and implementation.

Let be $G(t)$ a connected non-directed graph on which a path search (P_{AB}) between starting node A and destination node B is going to be performed ($A,B \in N(t)$ and $A \neq B$) in $rt_{\max}$ response time. Let be F a food node, for which the proper odor coverage area S_F is already set. Let be β the type of β-tabu to be applied.

Since there will be several breeds of ants, it is necessary to differentiate their notation. Let be $\tau_{ij}^{X}(t)$ the trail of pheromone of colony X at link l_{ij} at time t, while ρ^X is the evaporation rate of colony X. Let be P_{XY} the best path found between nodes X and Y, and P_C the best path of the split colony to the intermediate node C. Let be $R_i(t)$ the sequence of links covered by each ant h_i till time t. Finally, let be α the total number of ants.

The proposal involves a centralized control, in charge of the initialization and the post processing, and the algorithm defining the behavior of any generic ant. On the one hand, the initialization phase of SoSACO involves setting up the graph, arranging each ant, and preparing global storage of best path currently found P_{XY} for the relevant pairs of nodes (AB, AF, and BF), as formalized next.

initialization	
step 0	$P_{AB} \equiv \emptyset$; $P_{AF} \equiv \emptyset$; $P_{BF} \equiv \emptyset$; $P_C \equiv \emptyset$; split:=FALSE;
step 1	$\tau_{ij}^{X}(0) = 0$, $\forall$ colony X, $\forall$ link l_{ij}
step 2	if B=F $\forall$ ant h_i **set starting node A, end node B, tabu(h_i)={A};** else if A=F $\forall$ ant h_i **set starting node B, end node A, tabu(h_i)={B};** else half := round(α/2); $\forall$ ant h_i „ i $\in$ **[1, half], set starting node A, end node B, tabu(h_i)={A};** $\forall$ ant h_j „ j $\in$ **[(half+1),α], set starting node B, end node A, tabu(h_j)={B};**

On the other hand, after setting up all the ants, each of them will start drifting across the graph. At any iteration, any ant will first check if it is at the destination (thus it will have found a solution). If not, it will check if the pheromone of the other breed is firstly found (first ant finding trails of the other colony), so it will have to split its own colony. If not, it will finally move to an adjacent node, checking then if any odor is present. In case it is, and provided that individual has not found odor trails before, it will have found a partial path (to the food node) and eventually a full solution (if the other complementary partial path is already available). The described behavior is formalized in the following algorithm:

ant i behavior

```
while current_time < rt_max do
    if current node = end node then
        /*prepare solution */
        prepare path (remove loops from R_i; if split colony, concatenate first half P_C);
        exploit solution (R_i);
        perform global update and evaporation (Eqs. (7.3) and (7.4) in Sect. 7.4);
        restart ant h_i (empty tabu list and set current node := starting node);

    else if the other breed's pheromone is firstly found then split colony (not split and τ^Y_ij (0) > 0)
        then split:=TRUE;
             prepare partial path (remove loops from R_i);
             store partial path P_C:= R_i;
             ∀ ant h_j of the same breed (starting node of h_j = starting node of h_i) do
                 if (j mod2)=0 then starting node of h_j := current node;
                     else end node of h_j := current node;
                 restart ant h_j;
    else
        /* make next move */
        choose next node n;
        include into R_i the link from current node to node n (l_mn);
        perform local update of pheromone (Eq. (7.2) in Sect. 7.4);
        current node := n;

        /* smell (check odor) */
        if odor firstly found ( O(F, current node) >0 and  O(F,k)=0 ∀ k ∈ tabu(h_i))
        then follow odor, obtaining partial path R' leading to node F;
             /*prepare partial solution */
             prepare path (remove loops from R_i and concatenate R_i');
             lock P_XF  (where X is the starting node);
             if first partial solution (P_XF ≡ ∅) OR is a better solution (C(P_XF)>C(R_i|R_i'))
                 then store it  P_XF := R_i|R_i';
                      if exists complementary solution (P_YF ≠ ∅),
                         then exploit solution (P_XF|P_YF);
             unlock P_XF ;
```

Notice that anytime an ant achieves a new solution, there is a simple way to exploit it (taking care of collisions between ants accessing the same resources):

exploit solution P
lock P_{AB} (prevent other ants from updating it); if first solution ($P_{AB} \equiv \varnothing$), then provide path P and store it $P_{AB} := P$;   else if better solution ($C(P_{AB}) > C(P)$) then store it $P_{AB} := P$; unlock P_{AB} ;

When choosing next nodes, ants prioritize the pheromone of the other breed. If not found nearby, they will seek odor traces (but only if they have not found odor before). Finally, if that is not the case, they will take any node with choice probability proportional to the trails of their own pheromone (if present; if not, a random choice is performed). Choices based on pheromone were defined in Eq. (7.1).

choose next node n	for ant h_i from node m, having start node X and end node Y

take nodes adjacent to node m as $J \equiv \{j,, \exists\, l_{mj} \in L(t)\}$;
if the other pheromone is present in accessible link ($\exists\, j \in J$ meeting $\tau^{Y}_{mj}(0) > 0$)
  then do stochastic choice of next node $n \in J$ proportional to pheromone Y (Eq. (7.1), Sect. 7.4);
  else if not odor individually found yet ($O(F,k)=0\ \forall\, k \in tabu(h_i)$)
    and odor is present in adjacent node ($\exists\, j \in J$ meeting $O(F,j)>0$)
    then choose next node $n \in J$ maximizing odor (not $\exists\, j \in J$ meeting $O(F,j) > O(F,n)$);
  else do stochastic choice of next node $n \in J$ proportional to pheromone X (Eq. (7.4), Sect. 7.4);	

When an ant detects odor trails, it is able of finding its way to the food node by following the increasingly strong traces of odor. The method is formalized next:

follow odor	from node m, returning partial path R'

$R' \equiv \varnothing$;
stepnode := m;
while stepnode $\neq$ F do
  take nodes adjacent to node m as $J \equiv \{j,, \exists\, l_{mj} \in L(t)\}$;
  choose next step $n \in J$ maximizing odor (not $\exists\, j \in J$ meeting $O(F,j) > O(F,n)$);
  include link $l_{stepnode,n}$ into R';	

Finally, after the service is completed, the best solution found is pervaded with radial food odor trails. Notice that, unlike the odor diffusion in preprocessing time, this process has no minimum odor boundary.

```
post processing          having best solution found P_AB

if solution passes through the food node F (l_IF ⊂ P_AB) then
  obtain subpaths P_AF and P_FB arranged as sequences of links departing from node F;
  for each subpath do
     current node n := F;
     while subpath is not empty do
        extract next link l_nx from subpath;
        calculate odor for next node x coming from current node n as O_n(F,x) = O(F,n) − k · w_nx
        if that new odor trail is greater than the currently stored O_n(F,x) > O(F,x)
           then update odor O(F,x) := O_n(F,x);
```

7.5 Summary and Challenges

This chapter has described a metaheuristic method of the ACO type aimed to provide fast response to services requesting paths between two nodes (source and target nodes) across a huge dynamic graph. The method has been presented through its evolution for being adapted to the problem characteristics. Its evaluation [40] shows that it is able of quickly providing a first response (much faster than other methods) and then refine it in time. While the quality of the solutions provided is not optimal, it is good enough and its agility avoids the expiration of the validity of the solution provided in most cases. Furthermore, in those cases in which the solution loses validity, it is able of quickly adapting to changes and of speedily producing a similar solution.

The method counts on a preprocessing (food essence) but is locality eases updating it whenever it loses validity. In addition, food essence trails are calculated and set when the server is idle (not during service time), thus avoiding being a hindrance for service processing. Anyhow, the method works without the support of the preprocessing, so it will be always working, even in the meanwhile from the moment the obsolete food trails are removed to the moment they are re-implanted again (it will work slower, but the service won't lose availability).

The method is set for handling large amount of data, for which the use of a DB is recommended. DB technologies not only satisfy that need, but also fulfills the need for a light communication mechanism (blackboard), and support sharing partial results.

However, there are many challenges to make evolve the method. On the one hand, it is subject to several parameters, which forces setting preliminary experimentation for each implantation, aimed to discover an appropriate combination suited to the specific environment. It would be of use to ease their finding or even to propose a self-learning mechanism for those parameters. On the other hand, analyzing the automatic discovery of the most appropriate food nodes for any given problem is also very useful and challenging.

References

1. Dorigo, M., Maniezzo, V., Colorni, A.: The Ant System: an Autocatalytic Optimizing Process, Technical Report No. 91–016 Revised. Politecnico di Milano, Milan (1991)
2. Dorigo, M., Stützle, T.: Ant Colony Optimization. MIT Press, Cambridge (2004)
3. Dijkstra, E.W.: A note on two problems in connexion with graphs. Numerische Mathematic. **1**(1), 269–271 (1959)
4. Bast, H., Funke, S., Matijevic, D., Sanders, P., Schultes, D.: In transit to constant shortest-path queries in road networks. In: Proceedings of WS on Algorithm Engineering and Experiments (2007)
5. Delling, D., Sanders, P., Schultes, D., Wagner D.: Highway hierarchies star. In: 9th DIMACS Implementation Challenge: Shortest Paths (Nov 2006)
6. Delling, D., Holzer, M., Müller, K., Schulz, F., Wagner, D.: High performance multi-level routing. In: Demetrescu, C., Goldberg, A.V., Johnson, D.S. (eds.) The Shortest Path Problem: 9th DIMACS Implementation Challenge, DIMACS Book, vol. 74, pp. 73–92. American Mathematical Society, Providence (2009)
7. Bast, H., Delling, D., Goldberg, A., Müller-Hannemann, M., Pajor, T., Sanders, P., Wagner, D., Werneck, R.: Route planning in transportation networks. MSR Technical Report. (2014)
8. Chan, E.P.F., Lim, H.: Optimization and evaluation of shortest path queries. Int. J. Very Large Data Bases. **16**(3), 343–369 (2007)
9. Chan, E.P.F., Zhang, J.: A fast unified optimal route query evaluation algorithm. In: Proceedings of the 16th ACM Conference on Conference on Information and Knowledge Management. pp. 371–380 (2007)
10. Yu, J.X., Cheng, J.: Graph reachability queries: a survey. In: Aggarwal, C., Wang, H. (eds.) Managing and mining graph data. Advances in Data Base Systems, vol. 40, pp. 181–215. (2010)
11. Sankaranarayanan, J., Samet, H.: Distance oracles for spatial networks. In: Proceedings of the 25th IEEE International Conference on Data Engineering, pp. 652–663. IEEE, Piscataway (2009)
12. Sankaranarayanan, J., Samet, H., Alborzi, H.: Path oracles for spatial networks. In: Proceedings of the 35th International Conference on Very Large Data Bases, pp. 1210–1221. Association for Computing Machinery, New York (2009)
13. Binh, H.T.T., Duong, N.T.: Heuristic and genetic algorithms for solving survivability problem in the design of last mile communication networks. Soft. Comput. **19**(9), 2619–2632 (2015)
14. Kleinberg, J.: The small-world phenomenon: An algorithmic perspective. In: Proceedings of the 32nd ACM Symposium on Theory of Computing, 2000. Also appears as Cornell Computer Science Technical Report 99-1776 (October 1999)
15. Liben-Nowell, D.: Wayfinding in social networks. In: Algorithms for Next Generation Networks. Computer Communications and Networks Series, pp. 435–456. Springer Science+Business Media, Berlin (2010)
16. Mogren, O., Sandberg, O., Verendel, V., Dubhashi, D.: Adaptive dynamics of realistic small world networks. In: Procs. of the European Conference on Complex Systems (2009)
17. Honiden, S., Houle, M.E., Sommer, C., Wolff, M.: Approximate shortest path queries using Voronoi duals. In: Transactions on Computational Science IX, Lecture Notes in Comp. Science 6290, pp. 28–53. Springer, Berlin (2010)
18. Mehlhorn, K.: A faster approximation algorithm for the Steiner problem in graphs. Inf. Process. Lett. **27**(3), 125–128 (1988)
19. Erwig, M.: The graph Voronoi diagram with applications. Networks. **36**(3), 156–163 (2000)
20. Aparicio, S., Villazón-Terrazas, J., Álvarez, G.: A model for scale-free networks: application to twitter. Entropy. **17**(8), 5848–5867 (2015)
21. Chakraborty, A., Manoj, B.S.: An efficient heuristics to realize near-optimal small-world networks. In: 21st National Conference on Communications, NCC 2015 (2015)

22. Ridge, E., Curry, E.: A roadmap of nature-inspired systems research and development. J. Multiagent and Grid Syst. **3**(1), 3–8 (2007)
23. Bullnheimer, B., Kotsis, G., Strauss, C.: An improved ant system algorithm for the vehicle routing problem. Ann. Oper. Res. **89**, 319–328 (1999)
24. Lee, C.Y., Lee, Z.J., Lin, S.W., Ying, K.C.: An enhanced ant colony optimization (EACO) applied to capacitated vehicle routing problem. Appl. Intell. **32**(1), 88–95 (2010)
25. Ma, G., Duan, H., Liu, S.: Improved ant colony algorithm for global optimal trajectory planning of UAV under complex environment. Int. J. Comput. Sci. Appl. **4**(3), 57–68 (2007)
26. Abdallah, H., Emara, H.M., Dorrah, H.T., Bahgat, A.: Using ant colony optimization algorithm for solving project management problems. Expert Syst. Appl. **36**(6), 10004–10015 (2009)
27. Myszkowski, P.B., Skowroński, M.E., Olech, Ł.P., Oślizło, K.: Hybrid ant colony optimization in solving multi-skill resource-constrained project scheduling problem. Soft. Comput. **19**(12), 3599–3619 (2015)
28. Azar, D., Vybihal, J.: An ant Colony optimization algorithm to improve software quality in prediction models: case of class stability. Inf. Softw. Technol. **53**(4), 388–393 (2011)
29. Ramos, G.N., Hatakeyama, Y., Dong, F., Hirota, K.: Hyperbox clustering with ant colony optimization (HACO) method and its application to medical risk profile recognition. Appl. Soft Comput. **9**(2), 632–640 (2009)
30. Manaskasemsak, B., Rungsawang, A.: Web spam detection using trust and distrust-based ant colony optimization learning. Int. J. Web Inf. Syst. **11**(2), 142–161 (2015)
31. Wang, P., Lin, H.-T., Wang, T.-S.: An improved ant colony system algorithm for solving the IP traceback problem. Inf. Sci. **326**, 172–187 (2016)
32. Ahmad, M., Srivastava, J.: An ant Colony optimization approach to expert identification in social networks. In: First International Workshop on Social Computing, Behavioral Modeling and Prediction Phoenix. Springer, Arizona (2008)
33. Jindal, V., Bedi, P.: Reducing travel time in VANETs with parallel implementation of MACO (modified ACO). Adv. Intell. Syst. Comput. **424**, 383–392 (2016)
34. Niu, Y., Wang, S., He, J., Xiao, J.: A novel membrane algorithm for capacitated vehicle routing problem. Soft. Comput. **19**(2), 471–482 (2014)
35. Brito, J., Martínez, F.J., Moreno, J.A., Verdegay, J.L.: An ACO hybrid metaheuristic for close-open vehicle routing problems with time windows and fuzzy constraints. Appl. Soft Comput. J. **32**, 154–163 (2015)
36. G. Senarclens de Grancy, , M. Reimann. Vehicle routing problems with time windows and multiple service workers: a systematic comparison between ACO and GRASP (2016) CEJOR, 24 (1), 29–48
37. Eaton, J., Yang, S., Mavrovouniotis, M.: Ant colony optimization with immigrants schemes for the dynamic railway junction rescheduling problem with multiple delays. Soft. Comput. **20**(8), 1–16 (2015)
38. Dorigo, M., Blum, C.: Ant colony optimization theory: a survey. Theor. Comput. Sci. **344**(2–3), 243–278 (2005)
39. Rivero, J., Cuadra, D., Calle, J., Isasi, P.: Using the ACO algorithm for path searches in social networks. Appl. Intell. **36**(4), 899–917 (2012)
40. Calle, J., Rivero, J., Cuadra, D., Isasi, P.: Extending ACO for fast path search in huge graphs and social networks. Expert Syst. Appl. **86**, 292–306 (2017). https://doi.org/10.1016/j.eswa.2017.05.066
41. Borgatti, S.P.: Centrality and network flow. Soc. Networks. **27**(1), 55–71 (2005)
42. Min-Joonge, L., Sunghee, C., Chin-Wan, C.: Efficient algorithms for updating betweenness centrality in fully dynamic graphs. Inf. Sci. **326**, 278–296 (2016)

Index

F. Ros, S. Guillaume (eds.), *Sampling Techniques for Supervised or Unsupervised Tasks*, Unsupervised and Semi-Supervised Learning,
https://doi.org/10.1007/978-3-030-29349-9

The manufacturer's authorised representative in the EU is Springer
Nature Customer Service Centre GmbH, Europaplatz 3, 69115 Heidelberg,
Germany. If you have any concerns regarding our products, please
contact ProductSafety@springernature.com

Printed and bound by CPI Group (UK) Ltd, Croydon, CR0 4YY
05/10/2026
02228145-0005